DOE/EM-0654, Rev. 3
September 2011

United States of America
Fourth National Report
for the
Joint Convention on the Safety of Spent Fuel Management and on the Safety of Radioactive Waste Management

U.S. Department of Energy

In Cooperation with the
U.S. Nuclear Regulatory Commission
U.S. Environmental Protection Agency
U.S. Department of State

ABSTRACT AND ACKNOWLEDGEMENT

The United States (U.S.) of America ratified the Joint Convention on the Safety of Spent Fuel Management and on the Safety of Radioactive Waste Management (Joint Convention) on April 9, 2003. The Joint Convention establishes an international peer review process among Contracting Parties and provides incentives for nations to take appropriate steps to bring their n uclear activities into compliance with general safety standards and practices. The U.S. participated in Review Meetings of the Contracting Parties to the Joint Convention in November 2003, May 2006, and May 2009 in Vienna, Austria. This Fourth Report, an update of the U.S. National Report prepared under the Joint Convention in September 2011, documents spent fuel and radioactive waste management safety in the U.S. under the terms of the Joint Convention. It was prepared by the U.S. Government for review by the Contracting Parties.

The U.S. complies with the terms of the Joint Convention. An extensive U.S. legal and regulatory structure ensures the safety of spent fuel and radioactive waste management. The report describes radioactive waste management in the U.S. in both commercial and government sectors, and provides annexes with information on spent fuel and waste management facilities, inventories, and ongoing decommissioning projects. It also provides detailed information on spent fuel and radioactive waste management safety, as well as transboundary movements (imports/exports) and disused sealed sources, as required by the Joint Convention.

The U.S. Department of Energy acknowledges the support and cooperation of the U.S. Environmental Protection Agency, U.S. Nuclear Regulatory Commission, and U.S. Department of State in preparation of this report through the Joint Convention Interagency Executive Steering Committee and Working Group. The information in this report was extracted from publicly available information sources, including regulations and internet web sites of these agencies.

iii

U.S. Fourth National Report-Joint Convention on the Safety of Spent Fuel Management and on the Safety of Radioactive Waste Management

Copies of this report are available from:

U.S. Department of Energy
Center for Environmental Management Information
Office of Environmental Management
1000 Independence Avenue, SW
Washington, DC 20585

Mr. Douglas Tonkay
U.S. Department of Energy
Mailstop EM-43/Cloverleaf Building
1000 Independence Avenue, SW
Washington, DC 20585
Email: douglas.tonkay@em.doe.gov

TABLE OF CONTENTS

v

U.S. Fourth National Report-Joint Convention on the Safety of Spent Fuel Management and on the Safety of Radioactive Waste Management

vi

U.S. Fourth National Report-Joint Convention on the Safety of Spent Fuel Management and on the Safety of Radioactive Waste Management

vii

U.S. Fourth National Report–Joint Convention on the Safety of Spent Fuel Management and on the Safety of Radioactive Waste Management

viii

U.S. Fourth National Report-Joint Convention on the Safety of Spent Fuel Management and on the Safety of Radioactive Waste Management

ix

U.S. Fourth National Report-Joint Convention on the Safety of Spent Fuel Management and on the Safety of Radioactive Waste Management

LIST OF TABLES, FIGURES, AND ANNEXES

x

U.S. Fourth National Report-Joint Convention on the Safety of Spent Fuel Management and on the Safety of Radioactive Waste Management

U.S. Fourth National Report-Joint Convention on the Safety of Spent Fuel Management and on the Safety of Radioactive Waste Management

A. INTRODUCTION

This Fourth United States of America (U.S.) National Report updates the Third Report published in October 2008, under the terms of the Joint Convention on the Safety of Spent Fuel Management and on the Safety of Radioactive Waste Management[1] (Joint Convention). This report reflects developments in the U.S. through June 2011.

A.1 Purpose and Structure

This report satisfies the requirements of the Joint Convention for reporting on the status of safety at spent fuel and radioactive waste management facilities within the U.S.[2]

This Department of Energy (DOE) report was prepared by a working group composed of staff from DOE and other agencies of the U.S. Government involved in international and domestic nuclear activities, including Department of State, U.S. Environmental Protection Agency (EPA), and Nuclear Regulatory Commission (NRC).

The report format and content follow guidelines as agreed by the Contracting Parties to the Joint Convention. Chapters and annexes (or appendices) in this report have the same titles as in these guidelines, facilitating review by other Contracting Parties. Table A-1 provides a cross-reference between the chapters in this report and the specific reporting provisions in the Joint Convention. Section A.5 and Chapter K provide a concise summary of important changes since the Third U.S. National Report.

Table A-1 Joint Convention Reporting Provisions

National Report Section	Joint Convention Section
A. Introduction	
B. Policies and Practices	Article 32, Paragraph 1
C. Scope of Application	Article 3
D. Inventories and Lists	Article 32, Paragraph 2
E. Legislative and Regulatory Systems	Article 18; Article 19; and Article 20
F. General Safety Provisions	Articles 21-26; Articles 4-9; Articles 11-16
G. Safety of Spent Fuel Management	Articles 4-10
H. Safety of Radioactive Waste Management	Articles 11-17
I. Transboundary Movement	Article 27
J. Disused Sealed Sources	Article 28
K. Planned Activities to Improve Safety	Multiple Articles
L. Annexes	Multiple Articles

Information in this report is derived from publicly available information sources. More detailed information can be found at the internet web sites listed in Table A-2. The internet references provided in this report were available to the public and accurate as of the publication date. These URLs may change over time or may no longer be active.

---

[1] International Atomic Energy Agency, Joint Convention on the Safety of Spent Fuel Management and on the Safety of Radioactive Waste Management, INFCIRC/546, December 24, 1997.

[2] The U.S. ratified the Joint Convention on April 9, 2003.

Table A-2 Key Sources of Information Available on the Internet

Code of Federal Regulations

Access to all regulations: http://www.gpoaccess.gov/cfr/index.html

Energy, Title 10: (Includes DOE and NRC regulations): http://www.access.gpo.gov/cgi-bin/cfrassemble.cgi?title=200510

Protection of the Environment, Title 40: http://www.access.gpo.gov/cgi-bin/cfrassemble.cgi?title=200840

U.S. Department of Energy

Homepage: http://www.energy.gov

Office Health, Safety and Security: http://hss.energy.gov

Office of Environmental Management: http://www.em.doe.gov

Office of Nuclear Energy: http://www.ne.doe.gov

Office of Legacy Management: http://www.lm.doe.gov/

Energy Information Administration: http://www.eia.doe.gov/fuelnuclear.html

National Nuclear Security Administration: http://www.nnsa.doe.gov

Orders and directives: http://www.directives.doe.gov/

Waste Isolation Pilot Plant: http://www.wipp.energy.gov

U.S. Nuclear Regulatory Commission

Homepage: http://www.nrc.gov/

Regulations: http://www.nrc.gov/reading-rm/doc-collections/cfr/

Regulatory guides: http://www.nrc.gov/reading-rm/doc-collections/reg-guides/

Statutes and legislation: http://www.nrc.gov/reading-rm/doc-collections/nuregs/staff/sr0980/

Radioactive waste: http://www.nrc.gov/waste.html

Nuclear materials: http://www.nrc.gov/materials.html

Nuclear decommissioning (reactor and materials): http://www.nrc.gov/about-nrc/regulatory/decommissioning.html

Sealed sources and Devices: http://www.nrc.gov/materials/miau/sealed-source.html

Spent fuel storage: http://www.nrc.gov/waste/spent-fuel-storage.html

NARM toolbox: http://nrc-stp.ornl.gov/narmtoolbox.html

High-level waste: http://www.nrc.gov/waste/high-level-waste.html

Export/import: http://www.nrc.gov/about-nrc/ip/export-import.html

Japan Nuclear Accident - Near-term Task Force Report - Available from: http://www.nrc.gov/japan/japan-info.html

U.S. Environmental Protection Agency

Homepage: http://www.epa.gov/

Laws: http://www.epa.gov/lawsregs/laws/

Regulations: http://www.epa.gov/lawsregs/regulations/

Office of Air and Radiation: http://www.epa.gov/oar

Office of Resource Conversation and Recovery (formerly Office of Solid Waste): http://www.epa.gov/osw/

Radiation Program: http://www.epa.gov/radiation/

Waste Isolation Pilot Plant Oversight: http://www.epa.gov/radiation/wipp/index.html

Table A-2 Key Sources of Information Available on the Internet
Yucca Mountain Standards: http://www.epa.gov/radiation/yucca/index.html
Other
U.S. Department of State, Bureau of International Security and Nonproliferation (ISN): http://www.state.gov/t/isn
U.S. Defense Nuclear Facilities Safety Board: http://www.dnfsb.gov/
National Academies: http://www.nationalacademies.org
National Council on Radiation Protection and Measurements: http://www.ncrponline.org/
U.S. Nuclear Waste Technical Review Board (NWTRB): http://www.nwtrb.gov/
Conference of Radiation Control Program Directors, Inc.: http://www.crcpd.org/
U.S. Customs and Border Protection: http://www.cbp.gov/
Department of Homeland Security: http://www.dhs.gov
U.S. Public Health Service: http://www.usphs.gov/
U.S. Army Corps of Engineers Formerly Utilized Sites Remedial Action Program: http://usace.army.mil/CECW/PID/Pages/cecwm_natprog.aspx
Organization of Agreement States: http://agreementstates.org/
Interagency Steering Committee on Radiation Standards (ISCORS): http://www.iscors.org/
Radiation Source Protection and Security Task Force Report: http://www.nrc.gov/security/byproduct/2010-task-force-report.pdf
Low-Level Radioactive Waste Forum: http://www.llwforum.org
Blue Ribbon Commission on America's Nuclear Future: http://www.brc.gov

A.2   Summary Results from the Previous Review

The Guidelines Regarding the Form and Structure of a National Report requires National Reports to contain conclusions from the discussion of the Contracting Party's National Report at the previous Review Meeting. The discussions and conclusions from the Third Review Meeting as well as the questions and comments from the other Contracting Parties are reflected in this Fourth National Report.

The previous review raised a number of questions on the management of Greater-than-Class C Low-Level Waste (GTCC LLW), access to Class B and C low-level waste (LLW) disposal, and storage and disposal of high-level waste (HLW), and spent fuel. The alternatives for GTCC LLW disposal were of interest, especially whether these alternatives were keyed to the activated metals waste inventory in contrast to the immediate need for disposal of sealed sources. There was some interest in the role of the U.S. Congress in the selection of GTCC LLW disposal alternatives.

Decommissioning initiatives, especially complex facilities, were discussed in the country group review. Some queried whether the entombment (ENTOMB) option was used in the U.S. for nuclear power plant decommissioning. Currently, ENTOMB is not an option for decommissioning commercial nuclear power plants. A number of questions were raised on disposition of HLW, the status of the (former) Global Nuclear Energy Partnership, the lack of a national clearance standard and access for LLW disposal. Clarification of the U.S. approach for handling risk management issues and safety metrics for determining safety and operational trends was requested. Transparency was discussed, especially in cases where self-regulation applied in the U.S. waste management program. Challenges consisted of the continued need for permanent disposal of spent fuel and HLW, as well as to provide for GTCC LLW disposal. Access for disposal for LLW Class B and C was also recognized as an area for improvement.

Among the good practices mentioned about the U.S. safety program was acknowledgement of its increased focus on domestic disused sealed source tracking, collection, and disposition. There was appreciation of the publication of NRC regulations to implement the National Source Tracking System (NSTS) to enhance controls for certain radioactive materials considered to be a higher concern from a safety and security viewpoint. This prompted interest in how to establish a National Source Tracking System. Public participation and involvement in the decision making process was also highlighted, especially for mechanisms such as advisory boards, public meetings and hearings, rulemaking, transparency, and information available on the internet. The Global Threat Reduction Initiative (GTRI) was also commended with specific emphasis on GTRI's international cooperation to remove and secure high-risk nuclear and radiological materials and equipment posing a threat to the international community. The U.S. program to repatriate research reactor spent fuel from foreign countries was again cited with special reference to low-enriched uranium (LEU) fuel conversion. As in previous reviews, the remote-handled transuranic waste disposal conducted since 2007 at the Waste Isolation Pilot Plant (WIPP) was seen as exemplary experience in implementing concrete actions to promote safety and security. Finally, the peer review commended the U.S. active support and promotion of the Joint Convention.

A.3   Issues Addressed in the Third Review Meeting Summary Report

As part of this report, the U.S. has addressed six specific topics identified from the Summary Report of the Third Review Meeting of the Contracting Parties. These topics have been routinely addressed in this and past National Reports, but for ease in assessing the U.S. safety program they are provided with internal report references below.

A.3.1 Development of a Comprehensive Regulatory Framework

The U.S. has one of the most mature and well-established regulatory frameworks for nuclear safety. Section E of this report provides the legal and regulatory infrastructure including regulations, guidance, directives, inspections, and other mechanisms to promote safety and provide for enforcement of safety provisions.

A.3.2 Effective Independence of the Regulatory Body

In the context of commercial, academic, and industrial applications of nuclear material, the NRC has been established by the U.S. Congress as an independent regulatory agency. DOE exercises regulatory authority as discussed in more detail in Section E.2.3.1, E.2.3.2, and E.2.3.5, over nuclear activities conducted by DOE or on its behalf except in instances where Congress has specifically provided such regulatory authority to NRC. As discussed further in the report, the regulatory body in the U.S. is decentralized with NRC, DOE, EPA and numerous States having regulatory authority with respect to nuclear materials and facilities. Section E of this report addresses this regulatory arrangement.

A.3.3 Implementing Strategies with Specific Milestones

This issue was identified in the context of new actions initiated or enhancing of existing actions to improve the safe management of spent fuel and radioactive waste. Section A.5 of the report addresses this specific issue.

A.3.4 Funding to Secure Waste Management

The waste generator is financially responsible for waste management. The financial assurance mechanisms such as the Nuclear Waste Fund, letters of credit, bonds, and other financial devices are discussed in Section F.2.3 in some detail.

A.3.5 Educating and Recruiting Competent Staff and Employees

The need to maintain corporate knowledge has long been a concern of the U.S. in specific technological areas such as the nuclear field. Additional information is provided in A.5.8. U.S. agencies have robust recruitment programs to address the core competencies and projected needs. For example, NRC's Nuclear Safety Professional Development Program recruits recent graduates with strong academic records in health physics, earth sciences, or engineering. Additional information is available from the NRC's public website.[3] The NRC also has requirements for qualification on the part of its licensees (See F.2.1), training for State officials (See E.2.1.5), and for its own personnel such as inspectors.

A.3.6 Geological Repositories for High-Level Waste

Developments in the U.S. are discussed in Section A.5.1.

A.4 U.S. Actions Following the Japan Nuclear Accident

The IAEA has already taken steps in its context to address the accident with additional meetings, such as the June 2011 Ministerial Conference, sessions during the IAEA General Conference in September 2011 and a planned Extraordinary Meeting of the Convention on Nuclear Safety in August 2012. Although nuclear power plant safety is not within the scope of the Joint Convention, the safe management of spent fuel is. On March 11, 2011, when an earthquake and tsunami struck Japan affecting several reactors at the Fukushima Daiichi site, NRC fully activated its 24-hour Emergency Operations Center to monitor and analyze events at the nuclear plants in Japan. NRC moved to support the Japanese government and took a number of actions to ensure protection of the American people's health and safety.

In addition to the direct response in assisting Japan about the accident, the U.S. also undertook a number of steps to address the safety of nuclear power plants in the U.S. NRC created an agency task force, made up of current senior managers and former NRC experts with relevant experience, to conduct both short- and long-term analysis of the lessons learned from the situation in Japan, and the results of their work will be incorporated into future safety management activities for nuclear facilities. In the context of the Joint Convention, the safety of spent fuel, spent fuel storage and decommissioning following the accident are of singular importance.

---

[3] See http://www.nrc.gov/about-nrc/employment/training.html.

The NRC has performed a short-term systematic review, examining all available information from Japan, to see if there are changes that should be made to programs and regulations to ensure continued protection of public health and safety. NRC has also tasked inspectors who are posted at every U.S. nuclear power plant to perform special inspections at each plant to support the task force's short-term effort, supplemented as necessary by experts from the agency's regional and headquarters offices. The task force's recommendations for improving the safety of both operating and new nuclear reactors address improvements in the NRC programs for the oversight of reactor safety, as well as proposing an implementation strategy[4] The NRC will continue to evaluate whether any additional action by U.S. plants, are called for, prior to completing an in-depth investigation of the information from events in Japan.

The longer-term review will help inform any permanent NRC regulatory changes determined to be necessary. NRC is holding public meetings on the status of NRC's response to the Japan earthquake.

In addition, the U.S. nuclear power industry has taken actions at each licensed reactor site to verify measures to mitigate conditions that result from severe natural events, loss of electric power, flooding, and the loss of equipment functions during seismic events.

NRC is participating in many international forums to work with regulatory counterparts in other countries to share information and discuss lessons learned and actions that are being taken.

For additional information on NRC actions on the Fukushima accident, please refer to the following link: http://www.nrc.gov/japan/japan-info.html.

DOE has also taken steps to evaluate Fukushima Daiichi implications for its nuclear facilities. The Secretary of Energy issued Safety Bulletin 2011-01, entitled *Events Beyond Design Safety Basis Analysis* on March 23, 2011, stating that "...consistent with the approach being taken to review commercial nuclear power reactors, it is prudent to evaluate facility vulnerabilities to beyond design basis events at Department of Energy (DOE) nuclear facilities and to ensure appropriate provisions are in place to address them." Safety Bulletin 2011-01 asked Category 1 and 2 nuclear facility[5] managers to:

- Review how beyond design basis events have been considered or analyzed in accordance with DOE's Nuclear Safety Regulation and any controls that have been put in place that could prevent or mitigate them;
- Discuss the ability to safely manage a total loss of power event including a loss of backup capabilities; and
- Confirm safety systems are being maintained in an operable condition in accordance with technical safety requirements.

Although there are significant differences between commercial power reactors and DOE facilities in design, type, and amounts of radioactive material, and sources of energy,

---

[4] Recommendations for Enhancing Reactor Safety in the 21st Century (July 12, 2011) available at: http://pbadupws.nrc.gov/docs/ML1118/ML111861807.pdf.
[5] The facility categories are addressed in http://www.hss.doe.gov/enforce/docs/std/DOE_STD_1027cn1.pdf.

which could cause the release of radioactive material, there are lessons to be learned from the accident at the power reactors in Japan. These include the benefit of evaluating and putting into place provisions to address beyond design basis events, evaluating the impact of simultaneous events at multiple facilities, a total loss of power, and the loss of critical site and regional infrastructure (roads, communication systems, etc.) during and after an event. Safety Bulletin 2011-01 also seeks to determine the extent to which these conditions have been already evaluated at DOE facilities and to see if more analysis is warranted. Furthermore, the actions in the bulletin will confirm the availability of key equipment for response to both design basis events and beyond design basis events. DOE's Office of Health, Safety and Security has collected and is now evaluating responses to Safety Bulletin 2011-01.

DOE's Deputy Secretary convened a Nuclear Safety Workshop on June 6-7, 2011, in Washington, DC to further examine responses to Safety Bulletin 2011-01, other available information, and preliminary lessons-learned from continuing review of the Japan event. Participants included senior DOE, Federal agency, national laboratory, and industry executives. The workshop was very successful and was an important step as DOE moves forward with lessons learned and actions to improve nuclear safety at DOE.[6]

A.5 What is New Since Last Report

During the review meeting, a summary matrix was prepared for each Contracting Party by the country group rapporteur. To provide continuity from the Third Review Meeting, the U.S. rapporteur's matrix has been revised with citations to explanatory sections of the National Report to facilitate the review. Table A-3 presents the revised matrix with an overview of the U.S. program.

The U.S. has continued to increase its investment in the areas of nuclear and radiological security. Sections J.3 and K.3 describe the GTRI and other Interagency and international multilateral efforts (including coordination with the International Atomic Energy Agency (IAEA)). Through GTRI the U.S. continues to reduce and protect vulnerable nuclear and radiological material located at civilian sites worldwide through: (1) converting research reactors and isotope production facilities from the use of highly enriched uranium (HEU) to LEU; (2) removing and disposing excess nuclear and radiological materials; and (3) protecting high priority nuclear and radiological materials from theft and sabotage.

A number of challenges were identified at the Third Review Meeting. They are summarized in Table A-4 with their respective progress since the Third Review Meeting.

---

[6] The meeting agenda, attendance list, presentations, and other key workshop information are available at http://www.hss.doe.gov/nuclearsafety/ns/nsworkshop2011/index.html. A look back at the Nuclear Safety Workshop may be viewed at http://energy.gov/articles/look-back-nuclear-safety-workshop.

Table A-3 USA – Overview in Third Review Meeting Format

Type of Liability	Long-term Management Policy [7]	Funding of Liabilities	Current Practice/ Facilities	Future Facilities
Spent fuel	The Blue Ribbon Commission on America's Nuclear Future is evaluating alternative approaches and developing recommendations for management and disposal See A.5.1, B.3.1, G.7, K.1	Fee for electricity generated and sold is collected from utilities and deposited in the Nuclear Waste Fund to pay for disposal; Fund subject to annual Congressional appropriation See F.2.3.2	On-site and away from reactors wet & dry interim storage (private & government property) NRC integrated spent fuel regulatory strategy Acceptance of foreign research reactor fuel See B.3, C.1, D.1, G, Annex D-1	Awaiting a new strategy See A.5.1, B.3.1, D.1.2, G, K.1
Nuclear fuel cycle wastes (all LLW included in Non-Nuclear fuel cycle wastes for brevity)	HLW: The Blue Ribbon Commission on America's Nuclear Future (see above) Uranium & Thorium (U&Th) recovery sites: Near surface disposal See A.5.1, B.3.1, B.4.3, B.4.4, B.4.5, E.2.2.4, K.1	All: Producer pays U&Th recovery sites: Long Term Surveillance Fund Financial assurance required by license See F.2.3.2, F.2.3.3, H.3.5	HLW: Interim storage U&Th recovery sites: surface disposal locally See B.3.4, B.4.3, B.4.4, D.2.2.3, F.4.2.5, H.3, Annexes D-2, D-3A, D-3B	HLW: Awaiting a new strategy review U&Th recovery sites: additional license applications expected See A.5.1, B.3.1, B.4.4, D.2.2.3, K.1, K.4, Annex D-3B
Non-Nuclear fuel cycle wastes	Defense HLW: The Blue Ribbon Commission on America's Nuclear Future (see above) TRU waste: geologic disposal LLW: near surface disposal See A.5.1, A.5.2, A.5.3, A.5.4, A.5.6, B.4.1, B.4.2, K.1, K.2	All: Producer pays Defense HLW and TRU waste: public funds LLW: licensees required to demonstrate financial qualifications See F.2.3.1,	Defense HLW: interim storage Defense TRU waste: disposal at WIPP LLW: 3 commercial sites Interim storage of GTCC LLW See B.4.2, D.2.1, D.2.2.1, D.2.2.2, H.1, K.2, K.3, K.4, Annex D-2	Defense HLW: Awaiting a new strategy review. Waste Solidification Building for certain waste from MOX Fuel fabrication LLW: 1 pending site for Class A, B, and C LLW disposal GTCC LLW disposal (environmental impact assessment in progress) See K.1, K.2, K.4, K.8

[7] Refer to LIST OF ACRONYMS AND ABBREVIATIONS at the end of the report.

U.S. Fourth National Report-Joint Convention on the Safety of Spent Fuel Management and on the Safety of Radioactive Waste Management

8

Table A-3 USA – Overview in Third Review Meeting Format

Type of Liability	Long-term Management Policy[7]	Funding of Liabilities	Current Practice/ Facilities	Future Facilities
Decommis sioning liabilities	Nuclear power plants (NPPs): Decontamination & Decommissioning (D&D) to be completed within 60 years Defense, milling and other sites: Based on risk See B.5, D.3, E.2.1.4, F.6, F.7.2, H.1.4	NPPs: D&D fund required by law Non-legacy[8] Sites: Producer pays Defense sites: Public funds for defense liabilities See F.2.3.4	Large number of facilities undergoing decommissioning/ remediation See D.3, E.2.1.4, F.6, Annexes D-4, D-5, D-6, D-7	Large number of facilities planned for decommissioning/ remediation See K.5, K.6, Annexes D-4, D-5, D-6, D-7
Disused Sealed Sources	Disposal or recycle See Section J	Licensee or governmental responsibility See Section J	Disposal at government & commercial disposal sites Interim storage of sources on site by licensees Global Threat Reduction Initiative: Off-site Source Recovery Project See Section I.1, J, K.3	GTCC LLW disposal (environmental impact assessment in progress) Potential new commercial disposal site(s) See D.2.1.1, K.3

---

[8] Non-legacy sites are sites that have regulatory control and/or owned/controlled by a commercial or government entity

U.S. Fourth National Report- Joint Convention on the Safety of Spent Fuel Management and on the Safety of Radioactive Waste Management

Table A-4 Challenges for the U.S. in the Safety of Spent Fuel and Radioactive Waste Management

Challenges	Current Status
The potential shortage of low-level waste (LLW) disposal access/capacity requiring additional storage solutions.	A range of activities to improve the LLW regulatory framework resulted from a strategic assessment of the commercial LLW program. These include updated guidance on extended storage, a proposed rulemaking on waste streams not originally considered in the development of Part 61 and other alternatives for disposal. Furthermore, the State of Texas issued a license for LLW disposal (excluding GTCC LLW) for a new commercial disposal site and construction is underway. Options for expanding access for radioactive sealed sources to commercial disposal pathways were discussed extensively by Federal, State, and local government stakeholders, waste compact and industry representatives, These discussions are ongoing.
The lack of a disposal facility for (GTCC) LLW	A draft Environmental Impact Statement (EIS) for the disposal GTCC LLW and DOE GTCC-like waste was issued in February 2011 for public comment and public hearings were subsequently held.
The lack of a national clearance standard and the impact to public confidence.	Although a national clearance standard would have regulatory benefits, it has been deferred because of higher priority tasks and limited resources. The current case-by-case decision process is fully protective of human health and safety.
Spent fuel and HLW disposal.	The Blue Ribbon Commission on America's Nuclear Future is evaluating alternative approaches concerning the back end of the fuel cycle and developing recommendations for management and disposal of spent fuel and HLW.

The following sections summarize progress made in several important areas since the previous report.

A.5.1 Spent Fuel and HLW Disposition

In 2009 the Administration announced that it had determined that developing a repository at Yucca Mountain, Nevada, is not a workable option and the Nation needs a different solution for nuclear waste disposal. The Secretary of Energy established a Blue Ribbon Commission on America's Nuclear Future in January 2010 to evaluate alternative approaches for managing spent fuel and HLW from commercial and defense activities. The DOE Office of Civilian Radioactive Waste Management (OCRWM) ceased to function on September 30, 2010. Related activities that were performed by OCRWM are now being performed elsewhere in DOE. DOE remains responsible for disposing of spent fuel and HLW.

10

U.S. Fourth National Report-Joint Convention on the Safety of Spent Fuel Management and on the Safety of Radioactive Waste Management

A.5.1.1 DOE Research and Development Activities for Spent Fuel and HLW

DOE is performing research and development (R&D) that will address critical scientific and technical issues associated with the long-term management of used nuclear fuel, including storage, transportation and disposal. Fuel-cycle alternatives will be studied within the components of separations alternatives, spent fuel disposition (evaluation of fuel degradation effects over long-time storage periods), and fuel cycle system evaluations (addressing the open, modified open, and closed fuel cycle options). DOE is also participating in international and bilateral activities, in order to provide the U.S. with an understanding of the fuel cycle activities of other countries and to leverage the expertise and technical assessments for different geologic media and waste forms. See Section K.7.

The main objective in this DOE R&D is to develop a suite of options that will enable future decision makers to make informed choices about how best to manage the spent fuel from reactors. This R&D will be performed on functions in storage, transportation, and disposal, including research in a variety of geologic environments. An additional objective is the demonstration of technologies necessary to allow commercial deployment of solutions for the sustainable management of spent fuel that is safe, economic, and secure.

DOE's Fuel Cycle Research and Development Program conducts science-based R&D that integrates theory, experiment, and high performance modeling simulation to develop the needed technologies. This program provides a more complete understanding of the underlying science supporting the development of advanced fuel cycle technologies and provides a sound basis for future decisions on the U.S. nuclear fuel cycle. Its mission is to develop alternatives to current commercial fuel cycle management strategies to enable the safe, secure, economic, and sustainable expansion of nuclear energy while minimizing proliferation risks by conducting research and development focused on nuclear fuel recycling and waste management to meet U.S. needs. The main objectives include development of options for spent fuel management, enhancement of overall nuclear fuel cycle proliferation resistance via improved technologies for spent fuel management, and continuation of improved fuel cycle economics and safety performance of the entire fuel cycle system.

All fuel cycle options likely will result in some amount of spent fuel and/or HLW that will require permanent disposal. The U.S. may collaborate with other countries to conduct joint experiments or data exchanges associated with underground research laboratories (URLs) and safeguards issues.

In addition, the U.S. will pursue storage and transportation R&D activities. Objectives are to identify, assess and prioritize R&D needs related to very long-term storage and transportation of high burn-up fuel, demonstration of spent fuel integrity for long-term storage (up to 300 years), retrievability, and transportation of spent fuel after very long-term storage. This effort will identify technical gaps that need to be addressed for the very long-term storage of nuclear fuel cycle materials. Technical issues associated with transportation of spent fuel and wastes will be identified and prioritized for each of three fuel cycle options – open, modified open, and closed. Security assessments for long-term storage would also be addressed with consideration of the pertinent technical and regulatory issues.

11

U.S. Fourth National Report-Joint Convention on the Safety of Spent Fuel Management and on the Safety of Radioactive Waste Management

Research is planned to evaluate fuel degradation effects over lengthy storage periods and the effect of marine environments on storage canisters and overpacks. Analysis and evaluations will include the advantages and disadvantages of centralized storage versus reactor site storage and evaluation of safeguards and security issues at storage sites. Studies will include evaluation of retrievability, transportability, transport security of fuel after long storage periods, and distributed versus centralized storage coupled with eventual transport to a repository or recycling facility. Storage packaging configurations will be developed for potential scenarios in order to provide the maximum flexibility for disposition.

The Extended Storage Collaboration Program (ESCP) is a consortium of organizations coordinated by the Electric Power Research Institute (EPRI) to investigate aging effects and mitigation options for the extended storage of spent fuel, followed by transportation. In November 2009, EPRI convened a workshop of over 40 representatives of the nuclear industry, federal government, national laboratories, and suppliers of used fuel dry storage systems to discuss potential issues associated with extended dry storage of spent fuel, i.e., storage considerably beyond the term of current and recently proposed NRC regulations.

The primary activity of the ESCP in 2010 has been significant progress toward completion of technical "gap analyses" conducted by three organizations: DOE; Nuclear Waste Technical Review Board (NWTRB); and NRC. The highest priority has been to identify additional information on extended spent fuel dry storage and subsequent transportation, including: long-term degradation of high burn up spent fuel cladding; corrosion of the exterior of the stainless steel, welded canisters containing the spent fuel in an inert atmosphere (e.g., helium) located in coastal marine environments; degradation of concrete used for shielding and structural purposes; and monitoring of both the internal canister condition and exterior canister environment. Discussions were held on the purposes, types, work required to start a potential long-term dry storage demonstration program at one or more facilities using high burn up spent fuel. DOE is taking the lead in this effort. TransNuclear (Areva) is taking the lead in 2011 to establish utility participation in the non-destructive examination of some of the welded canisters currently in service.

Meetings were held to elicit international participation in the program. Representatives from the U.S., Germany, Japan, United Kingdom, Spain, Korea, France, and Hungary attended one or more of these meetings. Over the next year, potential research, and identification of available suitable demonstration facilities in these countries will be pursued. The consortium is also working to identify all relevant information for extended storage that has already been produced worldwide.

A.5.1.2 <u>Current Status of Back End of the Nuclear Fuel Cycle</u>

The Blue Ribbon Commission on America's Nuclear Future is conducting a comprehensive review of policies for managing the back end of the nuclear fuel cycle. It will also provide recommendations for "...developing a safe long-term solution to managing the Nation's used nuclear fuel and nuclear waste." An interim draft report was issued in July 2012, and a final report will be submitted to the Secretary of Energy in January 2012[9] The conclusions in the final report and subsequent activities will be

---

[9] See http://www.BRC.gov.

12

U.S. Fourth National Report-Joint Convention on the Safety of Spent Fuel Management and on the Safety of Radioactive Waste Management

reported during the Fourth Review Meeting of the Parties of the Joint Convention (May 2012).

DOE filed a motion with an NRC Atomic Safety and Licensing Board (ASLB) on March 3, 2010, seeking permission to withdraw the license application for a high-level nuclear waste repository at Yucca Mountain. On June 29, 2010, the ASLB issued an Order denying DOE's motion to withdraw. This decision was appealed to the Comission. In October 2010, NRC commenced and continued with the orderly closure of Yucca Mountain License Application review activities. As of the end of June 2011, an appeal was pending before the Comission.

In response to the evolving national discussion on HLW and spent fuel management strategy, NRC has developed an integrated spent fuel regulatory strategy to:

- Evaluate technical and regulatory needs to support very long-term dry storage of spent fuel and HLW;
- Develop a regulatory framework for potential spent fuel reprocessing; and
- Assess regulatory needs for a broad range of disposal options.

Commercial reactor sites will continue to store spent fuel in reactor pools and in NRC-approved storage facilities.

NRC updated its Waste Confidence findings and rule[10] expressing NRC's confidence that spent fuel can be stored safely and without significant environmental impacts for at least 60 years beyond the licensed life of any reactor, and sufficient mined geologic repository capacity will be available when needed. This integrated approach to regulating the back-end of the fuel cycle will help NRC:

- Maintain HLW and spent fuel safety and security as national strategy changes;
- Develop information needed to support new and efficient regulations; and
- Leverage limited resources and preserve assets.

NRC is considering a rulemaking potentially updating the Waste Confidence rule to address the impacts of storage beyond a 120-year timeframe.

A.5.2 Commercial LLW Disposal

Challenges remain for commercial LLW, especially Class B and C LLW disposal. Many U.S. LLW generators do not have access to operating commercial disposal facilities. A joint effort is now underway to solve this problem. Waste generators, business communities, and local, state and Federal governments are actively pursuing alternatives.

Most Class B and C LLW (as well as Class A sealed sources) in 36 of the 50 U.S. states as well as U.S. territories have no disposal paths since the Barnwell disposal facility in South Carolina was closed to most U.S. waste generators on June 30, 2008. The disposal challenge is now more serious because the commercial uranium

---

[10] Available at 75 FR 81037; December 23, 2010. See http://www.gpoaccess.gov/fr/

13

U.S. Fourth National Report-Joint Convention on the Safety of Spent Fuel Management and on the Safety of Radioactive Waste Management

enrichment market is expanding in the U.S. and depleted uranium tailings will require disposal. DOE also has a significant depleted uranium inventory, which may be disposed in commercial facilities. Finally, 10 CFR Part 61 is nearly 30 years old, and there is increasing interest in updating it with new concepts and standards. NRC has proposed rulemaking for changes to 10 CFR Part 61 for site-specific analyses rulemaking for shallow land burial (see Section K.6). The following are some issues government, industry, and others are addressing:

A.5.2.1 Blending

U.S. industry is proposing alternative approaches for managing Class B and C LLW. One approach is blending higher activity LLW (Class B and C concentrations) with lower activity waste (Class A) to form a Class A mixture meeting waste acceptance criteria of a commercial facility accepting Class A waste from all U.S. states. The approach mainly involves mixing ion exchange resins and filter media, a significant Class B/C waste stream, at a commercial LLW processing facility. Blending is mixing LLW having different concentrations. It does not involve mixing radioactive waste with non-radioactive waste, which is considered to be dilution. The term blending is used in the context of waste disposal in a licensed facility – it does not apply in the context of releasing radioactivity to the general environment.

NRC regulations do not prohibit blending to lower the waste classification, nor do they explicitly address it. NRC published guidance in 1995 discouraging blending to lower waste classification in some circumstances, but acknowledging it is appropriate in other circumstances. Because of renewed interest, NRC is working with a variety of stakeholders to develop a range of options for NRC's position on LLW blending.

NRC adopted a new risk-informed and performance-based position on LLW blending in October 2010.[11] NRC has commenced a rulemaking to require operating and future disposal facilities conduct a site-specific performance assessment for disposal of low-level radioactive wastes. NRC is working closely with Agreement States to ensure their requirements are compatible. NRC also intends to exclude GTCC LLW from this guidance and to establish clear standards for blended waste homogeneity as well as criteria for performance assessment.

A.5.2.2 Depleted Uranium Disposal

Depleted uranium (DU) is a source material as defined by the Atomic Energy Act of 1954, as amended (AEA), and, if treated as a waste, would meet the definition of LLW. One of NRC's responsibilities is to ensure safe disposal of commercially generated LLW. When NRC regulations (10 CFR Part 61) on LLW disposal were developed, there were no commercial facilities generating significant quantities of depleted uranium waste. The impacts of disposal of large amounts of depleted uranium were not considered in the development of Part 61; as a result, depleted uranium is considered to be Class A LLW. NRC issued licenses for two commercial uranium enrichment facilities (one in 2006 and one in 2007), and is reviewing license applications for two more. These facilities could generate quantities of depleted uranium significantly larger than considered during the development of 10 CFR Part 61. In addition, DOE has a significant depleted uranium inventory and may consider disposal at commercial facilities. NRC is now amending is

---

[11] Staff Requirements-SECY-10-0043-Blending of Low-level Radioactive Waste, October 13, 2010.

14

U.S. Fourth National Report-Joint Convention on the Safety of Spent Fuel Management and on the Safety of Radioactive Waste Management

regulations to require a site-specific analysis for disposing of waste streams not considered in the development of Part 61, including depleted uranium. The rulemaking is expected to be completed in late 2012. See Section K.6.

A.5.2.3 New Disposal Capacity

Waste Control Specialists, located near Andrews, Texas, is soon expected to provide Class A, B, and C LLW disposal capacity to generators within the Texas Compact (Texas and Vermont). The site is privately owned and regulated by the State of Texas. Construction is in progress and operations are expected to begin in 2012. The Texas Compact is establishing its rates and rules for importing waste from other States or Compacts. Preliminary rules, adopted in early 2011, allow approval for waste disposal from generators outside the Texas Compact. In addition, Waste Control Specialists is licensed to build a separate facility for disposal of LLW and mixed LLW that is the responsibility of the Federal government under section 3(b)(1)(A) of the LLRWPAA. The disposal facility is regulated by the State of Texas.

A.5.2.4 Potential Changes to U.S. LLW Regulatory Framework

NRC is considering potential revisions to its 10 CFR Part 61 regulatory framework to make it more risk-informed and performance-based.[12] The regulation was originally promulgated in 1983. Since then, other approaches for defining LLW disposal standards have been developed, such as IAEA's General Safety Guide GSG-1. In addition, NRC's regulation considered waste streams being generated at the time the regulation was developed, and as noted earlier, there are new waste streams, such as depleted uranium from enrichment plants, not considered in the original technical basis. NRC is evaluating alternatives for revising 10 CFR Part 61 and is receiving stakeholder input on such changes and their impacts.

A.5.3 Disused Sealed Sources

The U.S. Government conducts a program to collect disused sealed sources from the commercial sector – that present public health and safety hazards or national security threats – for safe storage and eventual disposal. Section J describes the DOE/National Nuclear Security Administration (DOE/NNSA) Global Threat Reduction Initiative (GTRI) Off-Site Source Recovery Project (OSRP) to remove excess, unwanted, or orphaned radioactive sealed sources posing a potential risk to public health, safety, and/or national security. The initial scope of the project included only GTCC sealed sources.

However, since the September 11, 2001 attacks, the recovery mission has expanded to reflect broader public safety and national security considerations. In addition to disused GTCC sources, the expanded DOE/NNSA mission now includes recovery of a wide range of sources that, if designated as commercial waste, would be classified as Class A, B, C, and GTCC low-level radioactive waste. DOE/NNSA prioritizes recovery of registered disused radioactive sealed sources based on threat reduction criteria developed in coordination with NRC. DOE/NNSA and its partners have been able to recover more than 28000 sources having over 800000 curies from more than 1000 sites in the U.S., as well as over 1000 U.S. origin sources from abroad.

---

[12] See Section K.6 on site-specific analyses rulemaking for shallow land burial.

15

U.S. Fourth National Report-Joint Convention on the Safety of Spent Fuel Management and on the Safety of Radioactive Waste Management

In addition, NRC implemented regulatory changes strengthening domestic licensing requirements for the import and export of high-risk radioactive sources and materials. These revisions to 10 CFR Part 110 brought U.S. import/export controls in line with the revised IAEA Code of Conduct on the Safety and Security of Radioactive Sources and international import/export guidance. In 2010, Part 110 was revised to authorize the import of Category 1 and 2 sources under a general license.

The Energy Policy Act of 2005 established the Interagency Task Force on Radiation Source Protection and Security (Task Force). The Task Force made significant progress since its original 2006 report to the President and Congress on improving security of domestic radioactive sources.[13] The 2010 Radiation Source Protection and Security Task Force Report updates the status of radiation sources, addresses disposal paths and challenges, and identifies alternative technologies and progress in control and accountability of disused sources. The 2010 Task Force Report identified the most significant radiation source protection and security challenge as access to disposal for disused radioactive sources. Although this is a security initiative, the focus on obtaining a disposal path for these sources aligns with the safety goal of isolating them from the public and the environment. This report documents the success of programs such as the NSTS, Off-Site Source Recovery Project, Source Collection and Threat Reduction (SCATR) Program and the GTRI. It also chronicles the cooperation among the various Federal agencies, states, and local governments to further enhance radiation source security. Recommendations contained in the 2010 report will be accomplished by responsible entities in accordance with an implementation plan. Task Force representatives will continue to meet periodically to monitor progress in implementing recommendations.

A.5.4 Greater-Than-Class C LLW Disposal

Greater-than-Class C (GTCC) LLW includes sealed sources, activated metals, and other waste (contaminated debris) with radionuclide concentrations and/or half-lives exceeding NRC limits for near surface disposal as Class C LLW. These wastes may be classified by some Contracting Parties as intermediate level waste. Section 3(b)(1)(D) of the Low-Level Radioactive Waste Policy Amendments Act of 1985 (LLRWPAA, Public Law 99-240) assigned the Federal Government responsibility for the disposal of GTCC LLW that results from activities licensed by the U.S. NRC and Agreement States. The LLRWPAA also specified that such waste be disposed in a facility licensed by the NRC. There are no facilities currently licensed by NRC for disposal of GTCC LLW. As the Federal agency responsible for GTCC LLW disposal, DOE is analyzing alternatives for the disposal of GTCC LLW. DOE issued a Draft Environmental Impact Statement (EIS) for the Disposal of GTCC LLW and DOE GTCC-like waste in February 2011. The Energy Policy Act of 2005 requires DOE to submit a report on these alternatives to the U.S. Congress and to await Congressional action before making a final decision about which alternative(s) to implement. See Section D.2.1.1 for additional information.

---

[13] http://www.nrc.gov/security/byproduct/task-force.html.

16

U.S. Fourth National Report-Joint Convention on the Safety of Spent Fuel Management and on the Safety of Radioactive Waste Management

A.5.5 Accelerated Weapons-Usable Uranium Return to the U.S. and Russia

DOE/NNSA works in partnership with the IAEA, the Russian Federation, and other nations to remove and protect vulnerable nuclear material located at civilian sites worldwide. Accomplishments as of September 2010 include:

- Helped secure over 971 radiological sites around the world, containing over 1270 PBq (34 million Ci) of radioactivity;
- Provided assistance for return of approximately 880 kilograms of Russian-origin HEU spent fuel from civilian sites worldwide to Russia for secure storage and disposition;
- Removed over 1200 kilograms of US-origin HEU spent fuel from civilian sites worldwide for secure storage and disposition in the U.S.;
- Removed approximately 18 kilograms of non-US-origin HEU spent fuel from a civilian site overseas for secure storage and disposition in the U.S.; and
- Converted to LEU fuel or verified the shutdown of 72 HEU research reactors worldwide, including seven U.S. university HEU research reactors.

A.5.6 Waste Isolation Pilot Plant

DOE operates the Waste Isolation Pilot Plant (WIPP), a geologic repository, authorized for the disposal of transuranic (TRU) waste generated by atomic energy defense activities. WIPP is supported by programs that provide characterization, confirmation, and disposal for defense transuranic waste. WIPP now has over 12 years of safe operations with over 11 million highway truck miles traveled safely. As of December 2010, over 72000 cubic meters (m^3) of defense TRU waste has been disposed with over 9,200 shipments. In 2010, WIPP was recertified by the Environmental Protection Agency, and its hazardous waste facility permit was renewed by the State of New Mexico.

A.5.7 American Recovery and Reinvestment Act of 2009 Cleanup

DOE received $6 billion in American Recovery and Reinvestment Act funding to accelerate clean up of legacy radioactive waste at DOE sites. Some of these funds are being used to:

- Characterize, certify, ship, and dispose more than 8000 additional m^3 of TRU waste;
- Remove TRU waste from eight sites with small quantities of the waste;
- Ship and dispose an additional 1815000 metric tons of uranium mill tailings from the Moab, Utah, site to the Crescent Junction disposal site; and
- Permanently dispose an additional 100000 m^3 of low-level waste.

These activities are reducing the volume of waste stored at generator sites and accelerating their permanent disposal. The number of sites storing radioactive wastes has been reduced, resulting in a decrease in the potential risk of continued radioactive waste storage.

17

U.S. Fourth National Report-Joint Convention on the Safety of Spent Fuel Management and on the Safety of Radioactive Waste Management

A.5.8 Education and Training Initiative

The U.S. has numerous opportunities for education and training that pertain to Nuclear Energy in general, with safety of spent fuel and radioactive waste falling within this scope. The U.S. has initiatives to provide for centralized knowledge management informational tools to advance access for training and educational purposes. Opportunities exist for training and education of experts from other nations. Access to facilities, such as the Waste Isolation Pilot Plant (WIPP), would provide experts with field experience in a geologic repository setting.

In addition, DOE and NRC support nuclear R&D activities; human capital development activities such as faculty development grants,[14] curriculum development grants,[15] graduate fellowships, undergraduate scholarships, and; and infrastructure and equipment upgrades for university-based research reactors and laboratories. For example, the Nuclear Energy University Programs (NEUP) plans to issue new solicitations in support of these areas in 2012.[16] These initiatives to increase nuclear resources, capabilities, and workforce have broad implications for managing spent fuel and radioactive waste. NRC has programs for educating and recruiting competent staff and employees (See Section A.3.5), which recruits recent graduates with strong academic records in health physics, earth sciences, or engineering. Additionally there are over 40 utility/community college partnerships set up to educate and train a large cohort of technicians who would be eligible to enter the utility workforce. This program, called the "Nuclear Uniform Curriculum" is organized by the Nuclear Energy Institute (NEI) and is tied heavily to the Institute of Nuclear Power Operators (INPO) training requirements.[17]

Training the next generation of nuclear scientists, engineers, and technicians is a critical need for the continued safe operation of U.S. nuclear plants and as we move toward greater use of nuclear energy to meet our energy needs, address global climate change, and close the back end of the nuclear fuel cycle. The program's immediate objective is to attract qualified students to all disciplines related to nuclear energy such as nuclear, mechanical, chemical, and electrical engineering and technology; chemistry, health physics, materials science, radiochemistry, geology, instrumentation and control, and nuclear policy at universities and colleges located in the U.S.

The Fulbright Scholarship[18] Program is the flagship international educational exchange program sponsored by the U.S. government and is designed to increase mutual understanding between the people of the U.S. and the people of other countries. The Fulbright Program has provided participants – chosen for their academic merit and leadership potential – with the opportunity to study, teach and conduct research, exchange ideas and contribute to finding solutions to shared international concerns. The

---

[14] Funding Opportunity Number: HR-FN-0711-NED2
http://www.grants.gov/search/search.do;jsessionid=h9TnTpnD572w4hL1vJ2h1kVLDHTqmyGnzyG7CBSzb4sPKVvbMdtk!301391617?oppId=107333&mode=VIEW

[15] Funding Opportunity Number: HR-FN-0711-EDU6
http://www.grants.gov/search/search.do;jsessionid=h9TnTpnD572w4hL1vJ2h1kVLDHTqmyGnzyG7CBSzb4sPKVvbMdtk!301391617?oppId=107353&mode=VIEW

[16] http://www.nuclear.energy.gov/universityPrograms/neUniversity2a.html

[17] http://www.nei.org/careersandeducation/nuclear-uniform-curriculum-program/

[18] http://fulbright.state.gov/index.html

18

U.S. Fourth National Report-Joint Convention on the Safety of Spent Fuel Management and on the Safety of Radioactive Waste Management

Fulbright Program is sponsored by the U.S. Department of State's Bureau of Educational and Cultural Affairs and Public Affairs Officers in U.S. Embassies. The Fulbright Program offers a broad array of opportunities for studies including science, technology, and engineering awards and scholarships. These include opportunities for U.S. scientists and engineers to study abroad and international students to study in the U.S. Historically the Fulbright Program has not focused on spent fuel and radioactive waste management, but a new emphasis is being placed on opportunities in the nuclear energy sciences, including managing spent fuel and radioactive waste.

19

U.S. Fourth National Report-Joint Convention on the Safety of Spent Fuel Management and on the Safety of Radioactive Waste Management

B. POLICIES AND PRACTICES

This section summarizes U.S. policies and practices for spent fuel and radioactive waste management, and related nuclear activities. The Federal government is responsible for the safe disposal of spent fuel and HLW radioactive waste.

B.1 U.S. National Nuclear Activities Policy

The U.S. Government promotes the development of commercial nuclear power and nuclear technology for beneficial uses in medicine, industry, and research. The promotional and regulatory duties for commercial activities are assigned to different agencies.

The Nuclear Regulatory Commission (NRC) is an independent agency authorized to regulate private sector and certain government nuclear facilities, regulating the possession and use of nuclear materials as well as the siting, construction, and operation of nuclear facilities. It performs its mission by issuing regulations, licensing commercial nuclear reactor construction and operation, licensing the possession of and use of nuclear materials and wastes, safeguarding nuclear materials and facilities from theft and radiological sabotage, inspecting nuclear facilities, and enforcing regulations. NRC regulates commercial nuclear fuel cycle materials and facilities and commercial sealed sources, including disused sealed sources. Three types of commercial nuclear materials are regulated:

- **Special nuclear material** – ^{233}U or ^{235}U, enriched uranium, or plutonium;
- **Source material** – natural uranium or thorium or depleted uranium not suitable for use as reactor fuel; and
- **Byproduct material** – certain radioactive materials produced by a nuclear reactor or as a waste product from uranium and thorium recovery process.[19]

NRC is also responsible for licensing commercial nuclear waste management facilities, independent spent fuel management facilities, and disposal facilities for high-level waste (HLW) and spent fuel. NRC also oversees certain state programs where NRC has relinquished limited regulatory authority to the individual states.

The Department of Energy (DOE) has responsibility for, among other matters, nuclear energy, nuclear weapons programs, nuclear and radiological weapons nonproliferation, radioactive waste management, and new nuclear-related activities for environmental remediation of contaminated sites and surplus facilities. DOE has regulatory authority over its facilities and nuclear activities, and those operated or conducted on its behalf, except where NRC is specifically authorized by statute to regulate certain DOE facilities and activities.

The U.S. Environmental Protection Agency (EPA) establishes generally applicable environmental standards to protect the environment from hazardous materials and certain radioactive materials. EPA has authority to establish standards for remediating

---

[19] The AEA specifically defines byproduct materials of 4 types; only some of these materials fall under the Joint Convention scope. See Section B.2.3 for more information.

active and inactive uranium mill tailing sites, environmental standards for the uranium fuel cycle, and environmental radiation protection standards for management and disposal of spent fuel, HLW, and transuranic (TRU) waste. EPA promulgates standards for and certifies compliance at WIPP for disposal of defense-generated TRU waste. EPA standards, under the Clean Air Act, limit airborne emissions of radionuclides from DOE sites. The regulatory roles of the U.S. agencies for nuclear activities are described in detail in Section E.

B.2   Government and Commercial Entities

B.2.1   Commercial Sector

Owners and operators of nuclear power plants and other types of facilities generating radioactive waste manage the spent fuel and radioactive waste generated by their facilities prior to disposal. U.S. Federal or state governments regulate waste disposal sites. Government custody may occur at different stages of the waste management scheme depending on the type of radioactive waste and generating activity. The interdependencies between the steps in spent fuel and radioactive waste management are addressed in Section F.7.3. Section G and H provide additional information on commercial spent fuel and radioactive waste management; respectively. Decommissioning activities generate radioactive waste in both the commercial and government sectors. Section F.6 describes decommissioning activities.

B.2.2   Government Sector

DOE is responsible for and performs most of the spent fuel and radioactive waste management activities for Government-owned and generated waste and materials, mostly located on Government-owned sites. These activities include managing spent fuel remaining from decades of defense reactor operations, which ceased in the early 1990s. DOE has safely stored the remaining defense spent fuel and spent fuel generated in a number of research and test reactors since then. DOE also provides safe storage for the core of the decommissioned Fort St. Vrain gas-cooled reactor and the core of the Three-Mile-Island Unit 2 reactor damaged in a 1979 accident.

DOE has a system for managing government spent fuel and radioactive waste. This includes numerous storage facilities and processing facilities (treatment and conditioning). Operating disposal facilities for low-level waste (LLW) and transuranic (TRU) waste are further described in Section D.2.2 of this report. Other waste management treatment and disposal systems support cleanup and closure of facilities no longer serving a DOE mission. More information on spent fuel and radioactive waste facilities in the government sector is in Section D.

The U.S. also continues activities to remove and/or secure high-risk nuclear and radiological materials both domestically and internationally. Part of this initiative is continuing the program of accepting U.S. origin foreign research reactor spent fuel back into the U.S. for safekeeping and the recovery of disused sealed sources.

B.2.3 Spent Fuel and Radioactive Waste Classification

Regulations addressing various aspects of the generation and control of radioactive wastes and other nuclear activities are in the U.S. Code of Federal Regulations (CFR),

22

U.S. Fourth National Report-Joint Convention on the Safety of Spent Fuel Management and on the Safety of Radioactive Waste Management

specifically Title 10 (Energy) and Title 40 (Protection of the Environment) of the CFR. They address the storage, treatment, possession, use and disposal of spent fuel and radioactive waste. Section E discusses various regulations. The U.S. classification system has two separate subsystems. One classification subsystem applies to commercial waste and is defined in NRC regulations. The other classification subsystem applies to DOE spent fuel and waste. The two systems are used for different purposes and different situations so conflicts do not occur. If ownership of radioactive waste is transferred from DOE to a commercial entity licensed by NRC, the waste is then subject to NRC regulation (and classification).

B.2.3.1 Spent Fuel

The U.S. defines "spent fuel" as fuel that has been withdrawn from a nuclear reactor following irradiation, the constituent elements of which have not been separated by reprocessing.

B.2.3.2 Radioactive Waste

Radioactive wastes in the U.S. have many designations depending on their hazards and the circumstances and processes creating them. NRC regulates most, but not all, sources of radioactivity, including LLW and HLW disposal, and residues from the milling of uranium and thorium.[20] Uranium mill tailings, the final byproduct of uranium ore extraction, are considered radioactive wastes. Radioactivity can range from just above background to very high levels, such as parts from inside the reactor vessel in a nuclear power plant. The day-to-day trash generated in medical laboratories and hospitals, contaminated by medical radioisotopes, is also designated radioactive waste.

NRC regulations in 10 CFR Part 61 classify LLW in the commercial sector as Class A, Class B, and Class C.[21] Waste that exceeds the specific activity of Class C LLW (referred to as Greater-than-Class C LLW) is considered generally unacceptable for near surface disposal absent additional specific safety requirements. This classification is based on potential LLW hazards, and disposal and waste form requirements. Class A LLW contains lower concentrations of radioactive material than Class B LLW, which has lower concentrations than Class C LLW. Table B-1 provides the commercial waste classes.

Radioactive waste owned or generated by DOE is classified as HLW, TRU waste, or LLW. In addition, DOE manages large quantities of uranium mill tailings and residual radioactive material.[22] Waste may also contain hazardous waste constituents. Waste with both radioactive and hazardous constituents in the U.S. is called "mixed" waste (mixed LLW or mixed TRU waste). DOE considers spent fuel to be nuclear material, and not a waste. Generally, the source of HLW is reprocessed spent fuel. TRU waste generally consists of protective clothing, tools, glassware, equipment, soils, and sludge contaminated with man-made radioisotopes beyond or "heavier" than uranium on the periodic table of the elements (long-lived alpha-emitting waste with concentrations greater than 3700 Bq/g [100 nCi/g]).

---

[20] Referred to in Section 11(e).2 of the Atomic Energy Act as byproduct material.

[21] This classification system is primarily based on protection of the inadvertent intruder.

[22] This residual radioactive material was the result of the Manhattan Project and is managed under the UMTRCA Title I. See Section D.2.2.3.1.

Table B-1 U.S. Commercial Radioactive Waste Classification

Waste Class	Description
HLW	The highly radioactive material resulting from reprocessing of spent fuel, including liquid waste produced directly in reprocessing and any solid material derived from such liquid waste containing fission products in sufficient concentrations and other highly radioactive material that the NRC, consistent with existing law, determines by rule requires permanent isolation.[23]
Class A LLW	Class A waste is determined by characteristics listed in 10 CFR 55(a)(2)(i) and physical form requirements in 10 CFR 61.56(a). (U.S. does not have a minimum threshold for Class A waste).
Class B LLW	Waste that must meet more rigorous requirements on waste form than class A waste to ensure stability.
Class C LLW	Waste that not only must meet more rigorous requirements on waste form than Class B waste to ensure stability but also requires additional measures at the disposal facility to protect against inadvertent intrusion.
GTCC LLW	LLW not generally acceptable for near-surface disposal.
AEA Section 11e.(2) Byproduct Material	Tailings or wastes produced by the extraction or concentration of uranium or thorium from any ore processed primarily for its source material content, including discrete surface wastes resulting from uranium solution extraction processes. Underground ore bodies depleted by such solution extraction operations do not constitute "byproduct material" within this definition.[24]

B.2.3.3 Other Regulated Radioactive Materials

NRC regulates other radioactive materials, but does not designate them as waste in the context of the Atomic Energy Act (AEA) of 1954, as amended. The definition of byproduct material was expanded by the Energy Policy Act of 2005 (EPAct05) to include discrete sources of ^{226}Ra, other Naturally Occurring Radioactive Materials of similar hazard, and Accelerator Produced Radioactive Material. EPAct05 relates to "discrete" and not diffuse sources. The expanded definition includes material now defined as 11e.(3) and 11e.(4) byproduct material, which refers to the citation in the AEA.

 EPAct05 also allowed this newly defined material (not regulated as low-level radioactive waste) to be disposed of in either a licensed radioactive waste or a permitted non-radioactive waste disposal facility. NRC can relinquish its EPAct05 authority to individual states[25] to regulate these radioactive materials.[26] Individual states usually regulate the radioactive materials not regulated by NRC.

On August 27, 2010, the final rule amending 10 CFR Part 110 became effective which changed export and import provisions relating to nuclear equipment and material. As a result of the 2010 rule change, the term "U.S.-origin" was added in the first exclusion to the definition of radioactive waste to clarify that the exclusion only applies to sources of

---

[23] From the Nuclear Waste Policy Act of 1982, as amended.

[24] Title 10 CFR Part 40, *Domestic Licensing of Source Material* (Section 40.4).

[25] In this context, "states" within the U.S. are similar to provinces or departments indicating the next level of government below the Federal level.

[26] More information is available from NRC's NARM Toolbox at http://nrc-stp.ornl.gov/narmtoolbox.html.

U.S.-origin. These sources may include sources with U.S. origin material and sources or devices manufactured, assembled or distributed by a U.S. company from a licensed domestic facility. Therefore, disused sources originating in a country other than the U.S. are now considered to be radioactive waste and a specific license for import is required by NRC for commercial licensees.

The Office of Surface Mining of the U.S. Department of Interior and the individual states regulate uranium ore mining. If there are elevated levels of diffuse radium or other naturally occurring radioactive materials, then EPA and individual states have jurisdiction. Other extraction mining and refinement operations for metals, phosphates, etc., may concentrate naturally occurring radionuclides in these tailings materials. NRC specifically licenses some mineral extraction processes (not for nuclear content), because they incidentally result in the use, or concentration, of material above 0.05 percent by weight source material. Identified processors are required to obtain an NRC license.

B.3 Spent Fuel Management Practices

This subsection provides information on spent fuel storage and disposal practices in the U.S. U.S. law generally uses the term "spent nuclear fuel" rather than spent fuel, and DOE has begun using the term "used fuel" to acknowledge that in the future, the material may have residual value through recycling. For purposes of this report, used fuel is referred to as spent fuel in accordance with the Joint Convention terminology. Past reprocessing activities are also described.

B.3.1 Spent Fuel Storage

The U.S. produces spent fuel in commercial nuclear power plants and research reactors. Currently 104 licensed nuclear power reactors provide about 20 percent of U.S. electricity. Information on U.S. nuclear power reactors is provided in the Convention on Nuclear Safety U.S. National Report[27] All operating nuclear power reactors are storing spent fuel in NRC licensed on-site spent fuel pools and over half are storing spent fuel in NRC-licensed independent spent fuel storage installations (ISFSIs) located on-site (see Annex D-1D). Given the circumstances regarding Yucca Mountain and the work being performed by the Blue Ribbon Commission on America's Nuclear Future, the current spent fuel management approach in the U.S. continues to be onsite storage at the nuclear power plants where the spent fuel was generated or at ISFSIs until a national long-term strategy is developed.

Most nuclear power plants that have been decommissioned or are undergoing decommissioning also have spent fuel stored on site pending disposal. Most permanently-shutdown commercial nuclear power reactors currently have or are planning to have their spent fuel stored at on-site ISFSIs. NRC amended its regulations in 1990 allowing licensees to store spent fuel in NRC-certified dry storage casks at licensed power reactor sites. Dry storage systems were developed as the preferred alternative (versus new pool construction). Most spent fuel is loaded in canisters with inert gas and welded closed. The canisters are then placed in storage casks or vaults/bunkers. Some cask designs can be used for both storage and transportation.

---

[27] http://www-ns.iaea.org/conventions/nuclear-safety.asp.

Sections D.1.1 and G provide additional information on spent fuel storage. Spent fuel is also stored at several research reactor sites licensed by NRC.

Since the Third U.S. National Report, NRC has renewed the licenses for four ISFSIs and currently has a fifth under review. These were renewed for a 40-year term, extending the total storage duration authorized by NRC for 60 years. NRC determined that the licensees' aging management plans along with their surveillance activities were sufficient to ensure that the spent fuel can be safely stored and retrieved at the end of the 60-year storage period.[28]

While the Blue Ribbon Commission on America's Nuclear Future is currently studying the long-term management options for spent fuel, NRC is staying focused on its regulatory responsibilities, such as the safety and security for continued onsite storage of spent fuel. Current sites likely will be active for many decades to come, given license renewals and the potential for new nuclear power plants to be built at existing reactor sites. NRC has developed a plan, called – the Integrated Spent Fuel Management Plan that focuses on extended storage of spent fuel, reprocessing, and alternatives to disposal of HLW at Yucca Mountain.[29] NRC's integrated spent fuel regulatory strategy and the ongoing review of its spent fuel programs will help advance the safety and security of continued interim storage of spent fuel.

Spent fuel from both domestic and foreign research reactors, in addition to limited quantities of commercial spent fuel, is stored at facilities at the Savannah River Site and the Idaho National Laboratory prior to further disposition. DOE continues to receive spent fuel from foreign and domestic research reactors. The program for receipt of foreign research reactor spent fuel is planned to be completed in 2019. No date has been set for completing receipt of spent fuel from domestic research reactors. DOE also stores spent fuel from former defense production reactors. DOE's current policy and planning includes managing foreign research reactor spent fuel for 40 years or until ultimate disposition.

B.3.2 Spent Fuel Disposal

The Nuclear Waste Policy Act (NWPA) of 1982 establishes the Federal responsibility for the disposal of spent fuel and HLW. The NWPA assigns responsibilities for the disposal of spent fuel and HLW to three Federal agencies:

- DOE for developing permanent disposal capability for spent fuel and HLW;
- EPA for developing generally applicable environmental protection standards; and
- NRC for developing regulations to implement EPA standards, deciding whether or not to license construction, operation, decommissioning and closure of the repositories, and certifying packages used to transport spent fuel and HLW to the licensed repositories.

The NWPA, as amended in 1987 by the Nuclear Waste Policy Amendments Act, directed DOE to characterize a site at Yucca Mountain, Nevada, for its potential use as a

---

[28] Standard Review Plan for Renewal of Spent Fuel Dry Cask Storage System Licenses and Certificates of Compliance — Final Report (NUREG-1927), March 2011.

[29] See http://www.nrc.gov/reading-rm/doc-collections/commission/speeches/2010/s-10-028.pdf

deep geologic repository. However, in 2009, the Administration announced that it had determined that developing a repository at Yucca Mountain, Nevada, is not a workable option and the Nation needs a different solution for nuclear waste disposal. The Secretary of Energy established a Blue Ribbon Commission on America's Nuclear Future in January 2010 to evaluate alternative approaches for managing spent fuel and HLW from commercial and defense activities. The DOE Office of Civilian Radioactive Waste Management (OCRWM) ceased to function on September 30, 2010. Related activities that were performed by OCRWM are now being performed elsewhere in DOE. DOE remains responsible for disposing of spent fuel and HLW.

B.3.3 Waste Confidence Decision and Rule

In 1984 and updated in 1990, NRC made a generic determination, referred to as the Waste Confidence Rule (10 CFR 51.23), that, if necessary, spent fuel generated in any reactor can be stored safely and without significant environmental impacts for at least 30 years beyond the licensed life for operation of that reactor. The licensed life includes the period of any revised or renewed license. The storage can be at a spent fuel storage "basin," an onsite or offsite ISFSI or both.

Recently, NRC updated[30] its Waste Confidence decision and rule, to express its confidence that spent fuel can be stored safely and without significant environmental impacts for at least 60 years beyond the licensed life of any reactor. The updated rule also noted the NRC's view that sufficient mined geologic repository capacity will be available when necessary. NRC is also considering another update to the Waste Confidence rule to include storage of spent fuel for periods greater than 120 years.

NRC has issued a final rule that increases the term of performance for storage casks from 20 years to 40 years in 10 CFR Part 72. The regulations would also allow possible renewal of the terms for such certification.[31] See section G.1 for additional details.

B.3.4 Reprocessing in the U.S.

Commercial reprocessing, where plutonium, uranium, or both are recovered from spent fuel to be used again in a reactor, was abandoned in the U.S. in the 1970s because of nuclear proliferation concerns. Several reprocessing ventures were considered in the 1960s and early 1970s. General Electric Company planned construction of a commercial reprocessing facility near Morris, Illinois, in the late 1960s, but only the storage facility was completed and remains in operation today.

Nuclear Fuel Services operated a reprocessing facility at West Valley, New York, from 1966 to 1972. This facility processed 640 metric tons of heavy metal (MTHM) from government and commercial nuclear power plants, resulting in 2.3 million liters of liquid HLW. This was the only commercial reprocessing plant operated in the U.S. The U.S. declared a moratorium on domestic spent fuel reprocessing in 1977. The moratorium was rescinded in 1981, but commercial reprocessing never resumed.

---

[30] 75 FR 81037; December 23, 2010. See http://www.gpoaccess.gov/fr/

[31] More specific information about the licensing process for both wet and dry storage facilities can be found at http://www.nrc.gov/waste/spent-fuel-storage.html.

DOE's Office of Nuclear Energy (NE) recognizes that R&D of sustainable fuel cycles and waste management activities are important to support the expansion of nuclear energy. NE will research and develop nuclear fuel and waste management technologies that will enable a safe, secure, and economic fuel cycle. The R&D strategy will be to investigate the technical challenges that would be encountered in each of three potential methods, and perform R&D within each one:

- *Once Through* – Nuclear fuel makes a single pass through a reactor after which the spent fuel is removed, stored for some period of time, and then directly disposed in a geologic repository for long-term isolation from the environment. The spent fuel will not undergo any sort of treatment to alter the waste form prior to disposal in this approach, eliminating the need for separations technologies that may pose proliferation concerns.

- *Modified Open Cycle* – The goal of this approach is to develop fuel for use in reactors that can increase utilization of the fuel resource and reduce the quantity of actinides that would be disposed in spent fuel. This strategy is "modified" in that some limited separations and fuel processing technologies are applied to the light water reactor fuel to create fuels that enable the extraction of much more energy from the same mass of material and accomplish waste management goals.

- *Full Recycle* – In a full recycle strategy, all of the actinides important for waste management are recycled in thermal- or fast-spectrum systems to reduce the radiotoxicity of the waste placed in a geologic repository while more fully utilizing uranium resources. In a full recycle system, only those elements that are considered to be waste (primarily the fission products) are intended for disposal, not spent fuel. Implementing this system will require extensive use of separation technologies and the likely deployment of new reactors or other systems capable of transmuting actinides.

The R&D approach will be to understand what can be accomplished in each of these strategies and then to develop the promising technologies to maximize their potential.

B.4 Radioactive Waste Management Practices

Radioactive waste management practices are summarized in the following sections of this report.

B.4.1 Low-Level Waste

Commercial and government facilities exist for LLW processing, including treatment, conditioning, and disposal. Generators prepare LLW for shipment to licensed disposal facilities. Section D.2.2.2 provides additional information on facilities and inventories of LLW.

Commercial LLW disposal facilities are designed, constructed, and operated under licenses issued by either NRC or an Agreement State, based on NRC health and safety regulations governing waste disposal quantities, forms, and activity levels (10 CFR Part

28

U.S. Fourth National Report-Joint Convention on the Safety of Spent Fuel Management and on the Safety of Radioactive Waste Management

61, *Licensing Requirements for Land Disposal of Radioactive Waste*). See Sections E.2.1.3 and H.1.1 for additional information.

LLW is disposed in near surface facilities; i.e., a land disposal facility in which radioactive waste is disposed of in or within the upper 30 meters of the earth's surface. Currently, commercial generators of Class B and C wastes in 36 States do not have access to a disposal site for these wastes and these wastes are being stored pending a disposal pathway (see Section D.2.1.2.). GTCC LLW is stored until an adequate method of disposal is established by DOE. GTCC LLW is discussed further in Section D.2.1.1. See Section D.2.2.2 for more details.

DOE operates disposal facilities for LLW owned or generated by DOE. DOE also uses commercial LLW disposal sites in certain circumstances. These practices are in Section F and Section H.

B.4.2 Transuranic Waste

Transuranic waste is managed by DOE. Defense TRU waste is disposed in the WIPP geologic repository. Remote-handled TRU waste emits more radiation than contact-handled TRU waste and must be both handled and transported in shielded casks. Section D.2.2.1 provides information on TRU waste disposal.

B.4.3 High-Level Waste

High-level waste from commercial reprocessing activities was vitrified and is stored at the former reprocessing plant in West Valley, New York. Defense HLW is stored, managed, and treated at three DOE sites (Savannah River Site, Hanford Site, and Idaho Site).

B.4.4 Uranium Recovery

Uranium recovery is the extraction or concentration of uranium from any ore processed primarily for its source material content.[32] The uranium recovery processes result in wastes that typically contain relatively low concentrations of radioactive materials having long half-lives. The wastes, in both solid and liquid forms, are classified as 11e.(2) byproduct material in accordance with definitions in the AEA. See Table B-1.

Three types of uranium recovery facilities have operated, are currently operating, or plan to operate in the future within the U.S. These are conventional mills, heap leach facilities, and in situ recovery facilities. Conventional mills and heap leach facilities extract uranium from ore processed above ground and consequently generate large volumes of solid 11e.(2) byproduct material. This material is disposed of in licensed near surface impoundment(s) on the site of the processing facility or in an off-site waste disposal facility licensed to accept 11e.(2) byproduct material. In situ recovery facilities differ from the other two types by leaching uranium from ore bodies in the subsurface. Consequently, the predominant waste stream for in situ recovery facilities consists of liquid wastes generated during their operation (typically less than 200 megaliters per year). The liquid wastes are disposed of by deep disposal well injection, by evapotranspiration to the atmosphere through land application of partially treated liquid

---

[32] Similarly, thorium was also extracted or processed in the past.

29

U.S. Fourth National Report-Joint Convention on the Safety of Spent Fuel Management and on the Safety of Radioactive Waste Management

waste, or by evaporation to the atmosphere from man-made lined ponds. The volume of solid waste generated at an in situ recovery facility (including salts from the evaporation process) is relatively small (typically less than 1000 cubic meters per year) and is ultimately disposed of off-site at a waste disposal facility licensed to accept 11e.(2) byproduct material. The most common type of facility currently being operated is the In situ recovery facility.

Prior to the mid 1980s, the sole type of uranium recovery facility in the U.S. was the conventional mill. Many of those previously operating facilities were reclaimed or are in the process of remediating (decommissioning) waste resulting from extracting uranium. Because of near-surface impoundments, those properties (and heap leach facilities) will be subject to long-term care after closure through ownership by governmental bodies. In situ recovery facilities do not include on-site disposal impoundments and thus do not require long-term care after closure. More detailed information is available in section D.2.2.3.2. Further discussion of disposal practices is in Section D.2.2.3. Solid radioactive wastes from the UMTRCA Title I at the inactive milling sites for non-commercial uses are discussed in Section D.2.2.3.1.

B.4.5 Enrichment and Fuel Fabrication Facilities Waste

The product from uranium recovery facilities is processed to enrich the fissile content. Tailings containing depleted uranium are a byproduct of the enrichment process. Fuel manufacturing facilities fabricate nuclear fuel assemblies for light water reactors containing low enriched uranium. This activity includes receipt, possession, storage, and transfer of special nuclear material. Other licensed activities supporting fuel manufacturing include uranium storage, scrap recovery, waste disposal, and laboratory services. Radioactive waste from these processes, which varies in type and amount, is managed within the classes described in Table B-1, e.g., Class A LLW.

Depending on available quantities, long-term and short-term needs, and cost/benefit analysis of potential uses, depleted uranium (DU) could be a resource for variety of applications and uses, in which case it is considered source material. If DU is not a resource, NRC categorizes it as Class A LLW. When 10 CFR Part 61 was developed, the disposal of large quantities of DU was not anticipated. However, with the recent licensing of fuel enrichment facilities, which will produce large quantities of DU waste, NRC determined it appropriate to revisit the issue. Therefore, NRC is examining whether the disposal of large quantities of DU from enrichment plants warrants additional, site-specific disposal protections to ensure long-term safety. As an interim measure, NRC has issued interim guidance to States, which regulate the disposal of large quantities of DU[33].

DU is currently possessed and stored by DOE and private corporations (e.g., U.S. Enrichment Corporation). DOE manages a large stock of DU at two gaseous diffusion enrichment plants. This DU continues to be managed as source material available for reuse. If a decision is made that this material has no potential use, it can be disposed in DOE or commercial low-level radioactive waste disposal facilities, provided the waste meets the disposal facility's waste acceptance requirements. Some DOE DU has been disposed as LLW at the Nevada National Security Site.

---

[33] See http://pbadupws.nrc.gov/docs/ML1002/ML100250501.html.

B.4.6 Ocean Disposal

The U.S. disposed of some LLW in the ocean in the 1950s and 1960s. This activity, while not specifically regulated, was an accepted method for managing low-level radioactive waste.[34] Ocean disposal of HLW was specifically prohibited. Ocean disposal of U.S. LLW was discontinued in 1970[35]

B.5 Decommissioning

Decommissioning generally happens at the end of operation of commercial and governmental nuclear facilities. Governmental agencies' recommendations, and in some cases requirements, include provision for decommissioning planning in the pre-operational design and strategy. Waste from decommissioning is managed within the waste classes in previously described. See Section F.6 for additional information.

Applicants for licenses are required to describe how facility design and procedures will facilitate eventual decommissioning[36] NRC has published regulatory guidance in Regulatory Guide 4.21, *Minimization of Contamination and Radioactive Waste Generation: Life Cycle Planning*, June 2008 for implementing this requirement.[37] The U.S. Governmental agencies work closely with industry, stakeholders, and members of the public to ensure lessons learned from decommissioning can be appropriately factored into the next generation of nuclear facilities (e.g., nuclear power plants, uranium mill facilities, and enrichment facilities).

---

[34] Radiation Protection at EPA: The First 30 Years. EPA 402-B-00-001, August 2000 at URL: http://www.epa.gov/radiation/docs/402-b-00-001.pdf.

[35] For additional information on the history and background of ocean disposal of radioactive waste, see History and Framework of Commercial Low-Level Radioactive Waste Management in the U.S.: ACNW White Paper, NUREG-1853, January 2007. http://www.nrc.gov/reading-rm/doc-collections/nuregs/staff/sr1853/sr1853.pdf.

[36] Regulations are stipulated in 10 CFR Part 20, Section 20.1406 of the U.S. Code of Regulations.

[37] See: http://www.nrc.gov/reading-rm/doc-collections/reg-guides/environmental-siting/active/ or from ADAMS accession number ML080500187.

C. SCOPE OF APPLICATION

This section covers the application of the Joint Convention in the U.S. (Article 3), the United States of America (U.S.) position on the application of the Joint Convention to reprocessing of spent fuel, naturally occurring radioactive material, and defense/military programs. This section also provides a definition of what the U.S. considers spent fuel and waste management facilities under the provisions of the Joint Convention.

C.1 Spent Fuel Reprocessing

The U.S. has no commercial reprocessing facilities. No declaration is therefore, needed under Article 3.1. If a decision is made in the future to proceed with construction of a reprocessing facility, the U.S. will make a declaration under Article 3.1 at that time.

C.2 Naturally Occurring Radioactive Materials

The U.S. does not consider naturally occurring radioactive materials (NORM) outside the nuclear fuel cycle to be within the scope of its Joint Convention obligations, as permitted by Article 3, paragraph 2. However, the Nuclear Regulatory Commission (NRC) has recently broadened the classes of byproduct material (designated as 11e.(3) and 11e.(4) byproduct material) to be regulated under the aegis of the Atomic Energy Act provisions. 11e.(3) and 11e.(4) byproduct material is regulated by NRC under 10 CFR Part 30. These new materials include discrete sources of ^{226}Ra and other NORM as well as accelerator produced material. Examples of this type of byproduct material produced by accelerators are: tritium (^{3}H), ^{14}C, ^{18}F, ^{87}Kr, and ^{57}Co. A discrete source is defined as a radionuclide that has been processed so that its concentration within a material has been purposely increased for use for commercial, medical, or research activities. Certain concentrations and quantities are exempt from the regulations.[38]

Diffuse sources of NORM include those that are not in the definition of discrete source. Furthermore, NRC does not regulate as discrete sources the inadvertent movement or concentration of NORM such as scale from pipes used in the fossil fuel industry, fly ash from coal power plants, or phosphate fertilizers.

C.3 Defense Activities

The Joint Convention does not apply to the safety of spent fuel or waste within defense or military programs unless declared specifically (Article 3.3). The U.S. Government has determined the Joint Convention does not apply to spent fuel or waste managed within the military programs in the U.S., but spent fuel and radioactive waste from military programs fall within the Joint Convention when transferred for permanent disposal in facilities operated by the Department of Energy (DOE).

U.S. military programs are primarily in the U.S. Department of Defense and the National Nuclear Security Administration (NNSA). NNSA is a separately organized agency within DOE, overseeing the military application of nuclear energy; maintaining and enhancing

---

[38] More specific information on discrete NORM sources can be accessed at http://www.nrc.gov/materials/byproduct-mat.html.

33

U.S. Fourth National Report-Joint Convention on the Safety of Spent Fuel Management and on the Safety of Radioactive Waste Management

the safety, security, reliability, and performance of the U.S. nuclear weapons stockpile; improving nuclear security through its defense nuclear nonproliferation programs; and development of naval propulsion plants for the U.S. Navy, among other functions.

The amount of spent fuel and radioactive waste from military programs is relatively small compared to the commercial nuclear power sector. Spent fuel and waste in military programs are managed, however, in accordance with the objectives stated in Article 1 of the Joint Convention.

The Joint Convention applies when waste and spent fuel are permanently transferred to an exclusively civilian program. The Joint Convention will apply to naval reactor spent fuel when accepted for disposal.

C.4 Radioactive Waste and Spent Fuel Management Facilities

The Joint Convention defines radioactive waste management as all activities, including decommissioning, handling, pretreatment, treatment, conditioning, storage, and disposal excluding off-site transportation. The U.S. has both commercial and Government radioactive waste management facilities under the Joint Convention.

The Joint Convention defines storage as holding radioactive waste in a facility for containment, with the intention of retrieval. The U.S. does not consider facilities as radioactive waste storage facilities where, for a short period of time (less than a year), a waste generator collects radioactive waste for shipment or processing before sending it to a treatment or disposal facility. This excludes low-level waste inventories at nuclear power plants, hospitals, universities, research facilities, industries, etc., where radioactive waste is generated and shipped to disposal sites. These facilities are subject to the regulations under licenses to possess nuclear materials. All such facilities, though not reported, subscribe to the same objectives of Article 1 of the Joint Convention.

Article 3 of the Joint Convention allows Contracting Parties to declare facilities undergoing decommissioning as radioactive waste management facilities. The U.S. has some on-site disposal facilities for radioactive waste being generated during cleanup. This report further discusses ongoing decommissioning (including site remediation) activities in Section D.3 and F.6.

34

U.S. Fourth National Report-Joint Convention on the Safety of Spent Fuel Management and on the Safety of Radioactive Waste Management

D. INVENTORIES AND LISTS

Radioactive waste inventories reported in this section are classified according to the waste classification definitions described in Section B of this report.

D.1 Spent Fuel Management

Most United States of America (U.S.) commercial spent fuel will remain stored at nuclear power plants until a disposition path is identified. Some spent fuel is also being stored away from nuclear power plants. The Joint Convention also applies to the Department of Energy (DOE) Government spent fuel storage facilities, including those used to store foreign research reactor and U.S. research reactor spent fuel transferred to DOE. Spent fuel management practices are discussed in Sections F and G.

D.1.1 Spent Fuel Storage

Spent fuel storage cask designers and manufacturers must comply with the quality assurance (QA) requirements in 10 CFR Part 72 Subpart G. The Nuclear Regulatory Commission (NRC) inspects storage designers, manufacturers, and licensees to verify QA procedures comply with their approved QA plan, and fabrication and use is done according to their QA program. See Section F.3 for additional information.

There are two primary canister-based dry cask storage systems for spent fuel.[39] One spent fuel design involves placing canisters vertically or horizontally in a concrete vault used for radiation shielding and protection of the canister. The other design places canisters vertically on a concrete pad and uses both metal and concrete storage overpacks for radiation shielding and canister protection.[40]

Table D-1 summarizes the types and numbers of U.S. spent fuel storage facilities and complete lists of spent fuel storage facilities are provided in Annex D-1. Figure D-1 shows the location of independent spent fuel storage installations (ISFSI) and other spent fuel storage facilities.

Table D-1 Spent Fuel Storage Facilities				
Function	Number of Facilities[41]	Inventory	Units[42]	Annex
Government				
Wet Storage	8	34	MTHM	D-1A
Dry Storage[43]	7	2420	MTHM	D-1A
University Research Facilities				
Wet Storage	21	1042	kg U	D-1B
Dry Storage	0	0	kg U	D-1B

---

[39] http://www.nrc.gov/waste/spent-fuel-storage/diagram-typical-dry-cask-system.html.

[40] http://www.nrc.gov/waste/spent-fuel-storage/designs.html

[41] In some instances multiple facilities at a given installation are counted as a single facility such as in the case of shared storage pools or ISFSIs.

[42] MTHM = Metric tons of heavy metal.

[43] Includes NRC-licensed facilities at the DOE Idaho Site and Fort St. Vrain in Colorado.

35

U.S. Fourth National Report-Joint Convention on the Safety of Spent Fuel Management and on the Safety of Radioactive Waste Management

Table D-1 Spent Fuel Storage Facilities				
Function	Number of Facilities[41]	Inventory	Units[42]	Annex
Other Research and Nuclear Fuel Cycle Facilities				
Wet Storage	3	36	kg U	D-1C
Dry Storage	1	102	kg U	D-1C
On Site Storage at Nuclear Power Plants[44]				
Wet Storage	68	49067	MTHM	D-1D
Dry Storage	52	15357	MTHM	D-1D

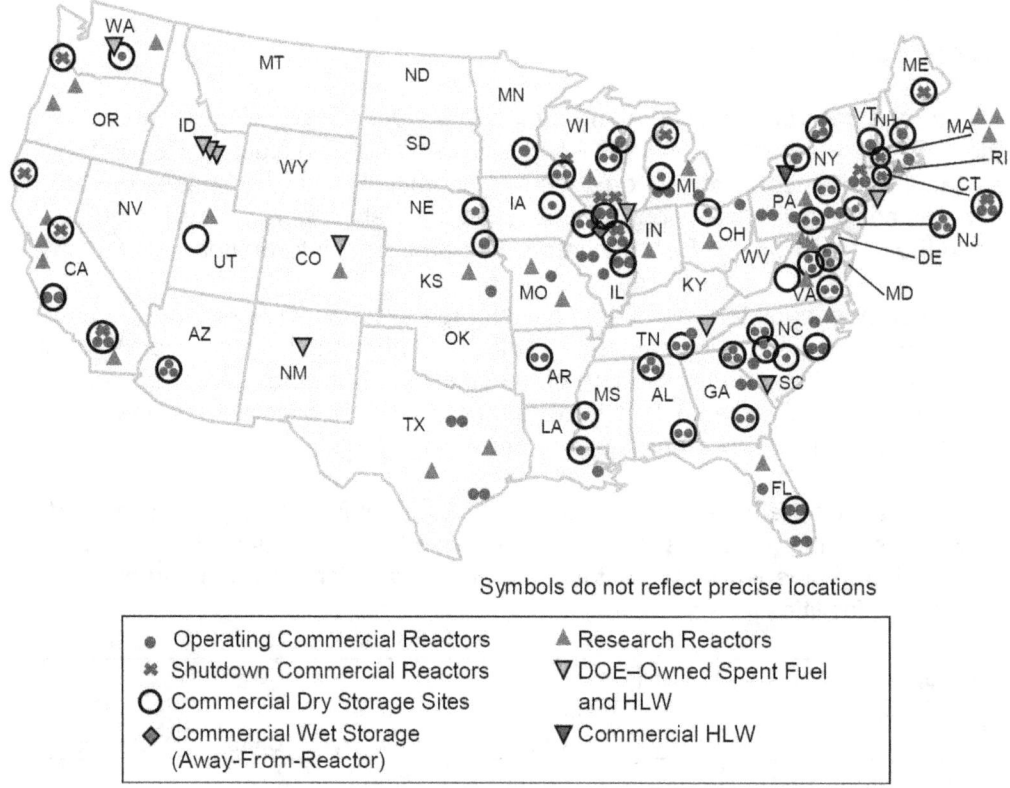

Symbols do not reflect precise locations

- • Operating Commercial Reactors
- ✱ Shutdown Commercial Reactors
- ◯ Commercial Dry Storage Sites
- ◆ Commercial Wet Storage (Away-From-Reactor)
- ▲ Research Reactors
- ▽ DOE–Owned Spent Fuel and HLW
- ▼ Commercial HLW

Figure D-1 Location of U.S. Spent Fuel and HLW Storage Installations

NRC regulations convey a general license to nuclear power reactor licensees to store spent fuel in dry storage systems approved by NRC at a site already licensed to operate or possess fuel for a nuclear power reactor under 10 Code of Federal Regulations (CFR) Part 50. NRC has already approved a variety of dry storage systems potential licensees may consider. These systems have Certificates of Compliance and are listed in NRC regulations (10 CFR 72.214). No applications or Safety Analysis Reports are required for a license to use these designs.

---

[44] Includes GE Morris and Utah Private Fuel Storage, which are not located at a nuclear power source.

The U.S. nuclear power industry generated 64424 metric tons heavy metal (MTHM) of spent fuel as of the end of 2010. Of this, 15357 MTHM is in dry storage at nuclear power plant sites. Also, 2454 MTHM of spent fuel is stored at government facilities. Projected inventories are updated annually.

D.1.2 Spent Fuel Disposal

The U.S. currently has no facility for spent fuel disposal.

D.2 Radioactive Waste Management

Section D.2.1 describes waste storage and treatment facilities and their associated inventories. Section D.2.2 describes disposal facilities in the U.S.

D.2.1 Radioactive Waste Storage and Treatment

Radioactive wastes are treated primarily to produce a structurally stable, final waste form and minimize the release of radioactive and hazardous components. The U.S. does not commonly make a distinction between the terms treatment and conditioning. Conditioning is defined in the international community as an operation producing a waste form suitable for handling, such as conversion of a liquid to a solid, enclosure of the waste in containers, or overpacking. Treatment is defined as those operations intended to improve the safety and/or economy by changing the characteristics of the waste through volume reduction, removal of radionuclides, and change in composition. U.S. terminology covering both conditioning and treatment is generally referred to as treatment or processing. Treatment is used in this broader context in this report.

Table D-2 summarizes the U.S. radioactive waste treatment and storage facilities and the inventory in storage.[45] Annex D-2 provides a list of facilities, their location, main purpose, and essential features. The following sections provide a brief description of the major types of radioactive waste management facilities.

Table D-2 Radioactive Waste Storage and Treatment Facilities

Sector	Function	Material Type	Number[46]	Inventory	Units	Annex
Government	Storage/ Treatment	HLW	7	340125	m^3	D-2A
		TRU	14	71417	m^3	D-2A
		LLW[47]	17	57571	m^3	D-2A
		AEA Section 11e.(2)	1	199000	m^3	D-2A
		Sealed Sources	2	2262	Containers/ Sources	D-2A
Commercial/ Other	Treatment/ Processing	LLW	64	Small volumes for collection		D-2B
	Storage	AEA Section 11e.(2)	1	21200	m^3	D-2B

---

[45] Stored inventories for LLW/MLLW and HLW are as of September 30, 2010. Stored inventories for TRU are as of December 31, 2009 per the Annual Transuranic Waste Inventory Report - 2010.
[46] In some instances, multiple facilities at a given installation are counted as a single facility.
[47] Includes mixed LLW.

D.2.1.1 Greater-Than-Class C LLW Management

Greater-than-Class C (GTCC LLW) waste is a form of low-level radioactive (LLW) waste containing long-and short-lived radionuclides with properties requiring a more robust disposal strategy[48] than for other classes of LLW. The authority to possess this type of radioactive material is included in NRC or Agreement State licenses. GTCC LLW may generally be grouped into the following three types: sealed sources, activated metals, and other waste. Other GTCC LLW includes contaminated equipment, trash, and scrap metal from miscellaneous industrial activities, such as manufacturing of sealed sources and laboratory research. Most GTCC LLW is activated metal, generated by decommissioning nuclear power plants, and disused sealed sources. Typical radionuclides associated with GTCC LLW are ^{14}C, ^{59}Ni, ^{94}Nb, ^{99}Tc, ^{55}Fe, ^{90}Sr, ^{238}Pu, ^{239}Pu, ^{241}Am, and ^{137}Cs. Although the U.S. inventory of GTCC LLW is modest, the construction of new commercial reactors and other proposed actions could generate additional quantities of GTCC LLW.

The Low-Level Radioactive Waste Policy Amendments Act of 1985 assigned the Federal government responsibility for disposal of GTCC LLW that results from NRC-licensed activities and specified that GTCC LLW be disposed in a NRC-licensed facility. There are no facilities currently licensed by NRC for the disposal of GTCC LLW. In addition, the Energy Policy Act of 2005 (EPAct05) requires DOE to complete several actions related to the preparation of an Environmental Impact Statement (EIS) and Record of Decision (ROD) for the disposal of GTCC LLW.

DOE is performing the National Environmental Policy Act analyses of potential GTCC LLW disposal alternatives, including developing an EIS. DOE published the Draft EIS (DEIS) for public review and comment in February 2011. The DEIS contains evaluations of various locations and technologies to optimize disposal of various types of GTCC LLW, as well as GTCC-like waste for which there is currently no disposal path. The DEIS considers alternatives for disposal in a geologic repository (WIPP), intermediate depth boreholes, enhanced near surface trenches, and above grade vaults. A preferred alternative may be a combination of alternatives. It also addresses candidate locations in various States and generic commercial disposal sites in four regions of the U.S. DOE held a series of nine hearings in calendar year 2011 for public review and comment.[49] DOE will consider stakeholder input when preparing the final EIS. DOE plans to issue a final EIS in 2012 and will submit a report to the U.S. Congress, as required by the EPAct05, and await Congressional action before making a final decision about which disposal alternative(s) to implement.

D.2.1.2 LLW Storage and Treatment

Commercial generators of LLW waste in the U.S. must treat these wastes to remove free liquids and stabilize or destroy other hazardous components contained in the waste. Wastes are also often treated to reduce the final disposal volume through compaction and incineration. Private companies in the U.S. provide processing (e.g., packaging and treatment) and brokerage services to facilitate safe storage, transportation and,

---

[48] In the context of the National Report, "more robust" means a greater degree of isolation, durability, and performance than is associated with near surface disposal for other classes of low-level radioactive wastes. This could include intermediate level waste as defined by some nations.

[49] See www.gtcceis.anl.gov.

38

U.S. Fourth National Report-Joint Convention on the Safety of Spent Fuel Management and on the Safety of Radioactive Waste Management

ultimately, disposal of LLW at one of three commercial disposal facilities. Some of these waste processor/brokers serve a limited clientele. Others perform these services for a wider body of clients. Annex D-2 includes a listing of such processors. Existing commercial LLW management facilities are regulated by Agreement States, in which they are sited; the Agreement States have put into place LLW regulations that are equivalent to those in the NRC's 10 CFR Part 61 regulations, as necessitated by law.

Many U.S. commercial generators of LLW could no longer dispose of their Class B and C low-level radioactive waste or their sealed sources since July 2008, when the Barnwell LLW disposal facility in the State of South Carolina limited its acceptance of waste to three U.S. states.[50] The Northwest Compact LLW disposal facility in the State of Washington limits its acceptance to 11 U.S. states. In anticipation of this circumstance, the NRC updated its guidance[51] related to extended interim storage of LLW for materials and fuel cycle licensees. For nuclear power licensees, NRC reviewed guidance prepared by the Nuclear Energy Institute, and concluded that the guidance was generally consistent with that of NRC (RIS-08-032).[52] NRC is being assisted by external subject matter experts in consolidating LLW storage guidance. The consolidated guidance may be available in late 2012.[53]

Because of disposal challenges, some in industry have proposed other solutions to mitigate the limitations in disposal access. For instance, one approach has been to thermally volume reduce and stabilize Class B and C waste to facilitate its storage.

Another of the approaches proposed by industry to deal with limited disposal access is blending of higher activity LLW (waste with Class B and Class C concentrations of radioactivity) with lower activity waste (Class A) to form a Class A mixture that can meet the waste acceptance criteria of an existing facility. In response to a recent NRC staff analysis and recommendation on the topic, the NRC, on October 13, 2010, issued Staff Requirements-SECY-10-0043. Section A.5.2 provides further details on this issue, as well as emerging NRC regulatory efforts to update in keeping with today's circumstances and state of knowledge.

NRC regulations do not prohibit blending to lower the waste classification, nor do they explicitly address it. NRC staff has published guidance that discourages blending to lower waste classification in some circumstances, but acknowledges that it is appropriate in others; e.g., 10 CFR 61.55(a)(8) provides that waste concentration may be averaged. The expanded blending proposed by industry has caused some stakeholders to question NRC's positions on blending. As a result, NRC is analyzing input from stakeholders and the public.[54]

---

[50] Connecticut, New Jersey, and South Carolina make up the Atlantic LLW Compact.

[51] RIS 08-12 *Considerations for Extended Interim Storage of Low-Level Radioactive Waste by Fuel Cycle and Materials Licensees*, May 2008. Available at: http://www.nrc.gov/reading-rm/doc-collections/gen-comm/reg-issues/2008/index.html.

[52] RIS 08-32. *Interim Low Level Radioactive Waste Storage at Reactor Sites*, December 2008. Available at: http://www.nrc.gov/reading-rm/doc-collections/gen-comm/reg-issues/2008/index.html.

[53] Commission paper SECY 10-0164, issued December 2010.

[54] Additional information on the decision support for LLW disposal is accessible at: http://www.nrc.gov/waste/llw-disposal/llw-pa.html.

D.2.1.3 Defense HLW and TRU Waste Treatment

Treatment facilities for defense waste are among the largest and most complex radioactive waste facilities in the U.S. Hundreds of millions of liters of HLW in tanks remain from decades of defense materials production activities.

DOE is building the world's largest radioactive waste treatment plant at the Hanford Site in southeastern Washington State to manage defense high-level waste stored for decades in 177 large underground tanks. The Waste Treatment and Immobilization Plant (WTP) will separate radioactive liquid waste and turn it into a stable glass form suitable for disposal. The radioactive liquid waste will be vitrified and poured into stainless steel canisters. The WTP will cover 26 hectares (65 acres) with four nuclear facilities – Pretreatment, Low-Activity Waste Vitrification, High-Level Waste Vitrification, and Analytical Laboratory – as well as operations and maintenance buildings, utilities and office space. Design of the plant will be complete by 2013; construction is scheduled to be completed just three years later, in 2016, and start-up of plant systems will begin. In 2019, all facilities and systems will be fully operational and begin the process of vitrifying tank waste.

The waste will first enter the Waste Treatment Plant complex in the Pretreatment Facility. The largest of the facilities in the complex, the Pretreatment Facility will separate the tank waste into low-activity and high-level waste streams for the Low-Activity Waste and High-Level Waste vitrification facilities. The Pretreatment Facility concentrates the waste by first removing any excess water. After that, the solids are filtered out using ultra-filtration technology and an ion exchange process removes the remaining soluble, highly radioactive material.

The Analytical Laboratory's key function is to ensure that the final, vitrified glass product meets all regulatory requirements and standards. When waste initially arrives at the WTP, the Analytical Laboratory will be used to determine the correct proportions of waste and glass forming materials that need to be mixed together. Determining the right "recipe" will produce a consistent vitrified waste product. The Analytical Laboratory will continue to sample waste throughout, making adjustments where needed so that all wastes will have consistent properties and meet accepted regulatory requirements.

The low-activity waste is primarily in liquid form that has a relatively small amount of radioactivity. These wastes will be sent to the Low-Activity Waste Vitrification Facility. After processing within this facility, the molten glass is poured into stainless steel containers. At the High-Level Waste Vitrification Facility, the vitrified HLW will be poured into stainless steel canisters.

Hanford's Integrated Disposal Facility has already been built for disposal of vitrified low-activity waste when operations begin. The Low-Activity Waste Vitrification Facility will produce approximately 1100 canisters per year. These canisters will be disposed at the Integrated Disposal Facility. HLW canisters will be temporarily stored at Hanford's Canister Storage Building until such time as a national repository for HLW is identified and constructed. About 480 canisters of HLW are expected to be produced each year during operations.

DOE is also constructing large treatment facilities for tank waste at the Idaho Site and Savannah River Site (SRS). At the Idaho Site, where most of the legacy tank waste has

40

U.S. Fourth National Report-Joint Convention on the Safety of Spent Fuel Management and on the Safety of Radioactive Waste Management

been treated and is stored as calcine, a new Sodium Bearing Waste Treatment Plant for treatment of remaining tank waste at the Idaho Site has been constructed and is expected to begin hot operations in 2012. This facility will treat the remaining tank waste at the Idaho Site, allowing closure of the four remaining underground tanks.

The Defense Waste Processing Facility located at SRS continues to vitrify tank waste. The Salt Waste Processing Facility is being constructed at SRS to replace existing treatment units. HLW recovered from the underground storage tanks will be processed in this new facility.

DOE/NNSA is constructing a Waste Solidification Building (WSB) at SRS, to process certain future waste streams from the Mixed Oxide Fuel Fabrication Facility and pit disassembly and conversion operations into a solid form for ultimate disposal. The WSB must be operational to support mixed oxide (MOX) cold start-up testing activities. The radioactive liquid waste consists of one high-activity (transuranic) and two low-activity streams. The high-activity stream contains americium removed from plutonium oxide during MOX aqueous polishing operations. The low-activity streams contain stripped uranium also removed from MOX aqueous polishing operations and laboratory waste from pit disassembly and conversion operations. The WSB operating life is expected to be approximately 15 years; however, the facility has a design life of 30 years.

In addition, treatment and certification of TRU waste continues at multiple DOE sites in preparation for disposal. Large treatment facilities are operating at the Idaho Site (Advanced Mixed Waste Treatment Facility) and Oak Ridge Reservation (Transuranic Waste Processing Center). Some of the other sites processing TRU waste are Argonne National Laboratory, Los Alamos National Laboratory, Hanford Site, and Savannah River Site. The historical inventory of TRU waste in storage continues to drop as waste is retrieved and processed for disposal.

D.2.2 Radioactive Waste Disposal

The cumulative inventory of disposed radioactive waste[55] is shown in Table D-3. Annex D-2 provides more detailed information on the quantities for each disposal facility.

D.2.2.1 Transuranic Waste Disposal

WIPP is a geologic repository to dispose, safely and permanently, TRU waste generated by atomic energy defense activities.[56] WIPP began operations on March 26, 1999, after more than 20 years of scientific study, public input, and regulatory review.

---

[55] Disposed inventories for LLW/MLLW and HLW in active facilities are as of 9/30/2010. Disposed TRU inventory (WIPP) is as of 12/31/2010. Commercial disposal inventories are as of 12/31/10.
[56] More information on WIPP can be found at http://www.wipp.energy.gov/index.htm.

41

U.S. Fourth National Report-Joint Convention on the Safety of Spent Fuel Management and on the Safety of Radioactive Waste Management

Table D-3 Radioactive Waste Disposal Facilities

Sector	Facility Type	Waste Type	Number	Inventory	Units	Annex
Government	Geologic Repository (WIPP)	TRU	1	72421	m³	D-2A
	Closed NNSS Greater Confinement Disposal (boreholes)	TRU	1	200	m³	D-2A
	Near Surface Disposal	LLW [57]	18	11859151	m³	D-2A
				122	Reactor Compartments	D-2A
Commercial	Operating Near Surface Disposal	LLW (Class A, B, C)	3	4413644	m³	D-2B
		AEA Section 11e.(2)	1	1345824	m³	D-2B
	Closed Near Surface Disposal	LLW	4	438143	m³	D-2B
Government/ Commercial	Title I UMTRCA Disposal	Residual Radioactive Material (tailings)	19	242450521	Dry Metric Tons [58]	D-3A
Commercial	Title II UMTRCA Disposal	AEA Section 11e.(2)	48			D-3A
Government	Other Closed Disposal Cells (Weldon Spring Site and Monticello)	Residual Radioactive Material (tailings)	2	3030000	m³	D-2A

WIPP is located in southeastern New Mexico, about 80 kilometers from Carlsbad. The repository consists of disposal rooms mined 655 meters underground in a 600-meter thick salt formation. This formation has been stable for more than 200 million years. WIPP-bound TRU waste is currently stored at multiple locations nationwide (see Annex D-2A). Over 72000 cubic meters (m³) of defense TRU waste was emplaced as of the end of calendar year 2010. The disposal limit, as defined in the WIPP Land Withdrawal Act, is 175675 m³.

WIPP has the capacity to accept all DOE defense contact-handled and remote-handled TRU waste in storage, and projected future waste generation. WIPP is authorized to only receive TRU waste from defense-related activities. DOE is currently assessing disposal alternatives, including potential disposal at WIPP, for other TRU waste.

---

[57] Includes Mixed LLW.
[58] Annex D-3A has additional quantities reported in units other than Dry Metric Tons.

D.2.2.2 LLW (Near Surface) Disposal

There are currently three active, licensed commercial LLW disposal sites. A fourth licensed site currently has facilities under construction.

- EnergySolutions/Chem-Nuclear, (Barnwell, South Carolina) — As of July 2008, access is limited to LLW generators within three states composing the Atlantic Compact (South Carolina, Connecticut, and New Jersey). Barnwell disposes of Class A, B, and C LLW up to 0.37 TBq (10 Ci) (which precludes many higher activity sealed sources).

- US Ecology (on DOE's Hanford Site near Richland, Washington) — restricted access to only the Northwest and Rocky Mountain Compacts. See Figure H.1 for states in these compacts. US Ecology disposes of Class A, B, and C LLW. The US Ecology site can also accept radium and other NORM and accelerator produced radioactive waste without compact restrictions.

- EnergySolutions (Clive, Utah) — accepts Class A LLW and mixed LLW for LLW generators without access to other compact facilities. EnergySolutions does not accept sealed sources.

- Waste Control Specialists (WCS) (near Andrews, Texas) — soon will provide Class A, B, and C LLW disposal to generators within the Texas Compact (Texas and Vermont). The site is privately owned and regulated by the State of Texas. Construction began in January 2011, and operations are expected to begin in 2012. The acceptance (import) of waste from out-of-Compact states for disposal is under consideration. On January 4, 2011, the Texas Compact Commission published rules to govern the acceptance (import) of LLW from other, non-compact waste generators. These rules allow petitions for access to the Texas Compact disposal facility for waste generators outside the Texas Compact. In addition, WCS is licensed to construct a separate facility for disposal of mixed LLW and LLW, designated a Federal responsibility under section 3(b)(1)(A) of the LLRWPAA. WCS can currently dispose of waste with exempt levels of radioactive material per Texas state law at its site.

Commercial LLW sites now closed are: Beatty, Nevada (closed 1993); Maxey Flats, Kentucky (closed 1977); Sheffield, Illinois (closed 1978), and West Valley, New York (closed 1975).

Table D-4 provides a breakdown of LLW commercially disposed in 2010, a representative year.[59]

Over 99 percent of the LLW volume disposed of at commercial sites was Class A LLW, most of which was disposed of at the Clive, Utah, site, with the remaining volume split between the Barnwell, South Carolina, and Richland, Washington, sites. Approximately 90 percent of the Class B LLW and 70 percent of the Class C LLW was disposed at the Barnwell site, with the remainder disposed at Richland.

---

[59] Source: Manifest Information Management System database, DOE December 2010, see http://mims.apps.em.doe.gov/.

Table D-4 Low-Level Waste Received at Commercial Disposal Sites in 2010 (Volume in m^3)

Source	Class A	Class B	Class C	Total
Low-Level Waste				
Academic	10	0	1	11
Government (from DOE)	73401	0	0	73401
Government (non-DOE)	21047	0	0	21047
Industry	33018	4	1	33023
Medical	12	0	1	13
Utility	5417	30	53	5500
All Other (Undefined)	1303	0	0	1303
Mixed Low-Level Waste				
Government	2222	0	0	2222
Non-Government	72	0	0	72
Total	*136502*	*34*	*56*	*136592*

It should be noted that there is no limit below which commercial low-level waste is outside of regulatory purview. However, NRC regulations' do provide alternative waste disposal approval processes that allow for the disposal of some low-activity low-level waste in landfills designed and regulated for other purposes (e.g., hazardous waste disposal or industrial debris disposal). Additional information on management strategies for these types of low activity wastes is provided in Section H.1.3.

D.2.2.3 Uranium Mill Tailings Disposal

Section B.4.4 describes uranium recovery facilities in the U.S. For conventional mills, the waste is primarily the on-site disposal of tailings (residual ore after the uranium was leached). The Uranium Mill Tailings Radiation Control Act (UMTRCA) classified the tailings either as residual radioactive material or 11e.(2) byproduct material depending on the status of the facility at the time UMTRCA was passed in 1978. UMTRCA Title I applies to facilities that were closed or abandoned prior to1978 and the waste at these sites is referred to as residual radioactive material. Since passage of UMTRCA, activities at Title I sites were largely focused on decommissioning and cleanup of residual radioactive material by the U.S. governmental entities. UMTRCA Title II applies to sites under an active license in or after 1978; waste at Title II sites is designated as AEA Section 11e.(2) byproduct material. See Tables B-1 and B-2.

D.2.2.3.1 UMTRCA Title I Mill Tailings Sites

UMTRCA Title I required DOE to complete surface remediation and groundwater cleanup at the listed inactive uranium milling sites at which the uranium was processed solely for sale to the U.S. government. Residual radioactive material, including any wind-blown dust, may have been consolidated into a single cell or perhaps relocated to a cell constructed on another site. These cells are now under long-term surveillance by DOE (or possibly state or Tribal governments in which the cell is located) and licensed by NRC under provisions in 10 CFR 40.27. Annual site inspections are performed as part of the long-term surveillance program at 22 Title I disposal sites. These sites are listed in Annex D-3A.

D.2.2.3.2 UMTRCA Title II Licensed Uranium Recovery Facilities

NRC and EPA established requirements and standards to be met prior to NRC termination of a license for a specific facility. The requirements include long-term stability of byproduct material disposal piles, radon emissions control, water quality protection and cleanup, and remediation of land and buildings. Once a license is terminated, ownership is transferred to a governmental entity for long-term care of the 11e.(2) byproduct material that was licensed for disposal at conventional mill or heap leach facilities under provisions in 10 CFR 40.28. For in situ facilities, no tailings are generated; no permanent surficial disposal options are licensed. Any remaining residues from evaporation ponds are collected and disposed of at an 11e.(2) byproduct material licensed off-site waste disposal facility. The subsurface aquifers that were subjected to the leaching process do not constitute 11e.(2) byproduct material and undergo restoration under 40 CFR Part 146 EPA Underground Injection Control (UIC) Program Criteria. Alternatively, liquid 11e.(2) byproduct material waste collected at the surface can be disposed of by deep well injection at the site of In situ operations, but such aquifers will be permanently restricted from future use by institutional controls.

Forty-eight UMTRCA Title II facilities have been licensed and consist of conventional uranium and thorium mills (27), heap leach facilities (2), in situ recovery (ISR) facilities (14), 11e.(2) byproduct material disposal facilities (2), a resin transfer facility, an oxide-hexafloride conversion facility and a former mining facility. Twenty-four facilities are located within and regulated by Agreement States. There are five Agreement States (Colorado, Illinois, Texas, Utah, and Washington) licensing AEA Section 11e.(2) byproduct material. The remaining 24 facilities are regulated by NRC and located within Nebraska, New Mexico, Oklahoma, South Dakota, or Wyoming.[60] Annex D-3A lists both – NRC and Agreement State – regulated uranium recovery facilities.

One conventional mill site license, the Maybell-West facility in Colorado, has been terminated in the past three years and the reclaimed tailings areas transferred to DOE for long-term care under the general license provisions of 10 CFR 40.28. NRC is required to determine applicable standards and requirements have been met before termination of the licenses at sites located in Agreement States.

Separate 11e.(2) byproduct material disposal cells (licensed separately from the LLW disposal cells) are: the EnergySolutions facility in Clive, Utah, which is regulated by Utah under Agreement State authority, and the WCS facility in Andrews County, Texas, which is regulated by Texas under Agreement State authority. Both facilities are listed under the Radioactive Waste Management Facilities (Annex D-2B).

D.2.2.3.3 Mine Overburden Remediation

Mine overburden is not classified as radioactive waste requiring restricted disposal, but an estimate of mine overburden is provided at the request of other Contracting Parties to

---

[60] The WNI Sherwood site in the state of Washington is regulated under NRC general license to DOE for long-term surveillance.

the Joint Convention.[61] Although there are about 4000 mines with documented production, a database compiled by EPA with information from other Federal, state, and Tribal governmental lands, includes 15000 mine locations, mostly in 14 western states.[62] Most of these locations are in Colorado, Utah, New Mexico, Arizona, and Wyoming, with about 75 percent of those on Federal and tribal lands. The majority of these sites were conventional (open pit and underground) mines. With the drop in market price of uranium beginning in the 1980s, U.S. producers turned to in situ recovery operations as a principal means of extracting uranium from ore bodies. There were 20 uranium mines operating in 2009 according to DOE's Energy Information Administration.[63]

Mining of uranium ores by surface and underground methods produces large amounts of radioactive waste material classified as naturally occurring radioactive materials (NORM) or Technologically Enhanced NORM (TENORM), including overburden, un-reclaimed sub-economic ores (protore),[64] "barren" rock, and drill cuttings. The volume of waste produced by surface, open-pit mining is a factor of approximately 45 times greater than for underground mining, based on their respective averages. Thus, the amount of overburden generated from open-pit mines far exceeds underground mine overburden. The U.S. Geological Survey in an estimate for U.S. Environmental Protection Agency (EPA), found the amount of waste rock generated by approximately 4000 conventional mines in their data files ranged from one billion to nine billion metric tons of waste, with a likely estimate of three billion metric tons.[65] Given the larger number of mine locations identified by EPA, the amount of waste rock is likely to be higher. Additional historical information is provided in the Third National Report.

D.3 Nuclear Facility Decommissioning

Table D-5 summarizes ongoing U.S. decommissioning activities within the Joint Convention. More information is provided in the following subsections corresponding to each of the entries in Table D-5.

---

[61] Unless otherwise noted, this information can be found at http://www.epa.gov/radiation/tenorm/uranium.html.

[62] U.S. Environmental Protection Agency, *Uranium Location Database Compilation*, EPA 402-R-05-009, August 2006.

[63] U.S. Energy Information Administration, 2010. *Domestic Uranium Production Report* (2003-2009), U. S. Uranium Mine Production and Number of Mines and Sources, 2003-2009. See http://www.eia.doe.gov/cneaf/nuclear/dupr/umine.html.

[64] Protore is material containing uranium that cannot be produced at a profit under existing conditions but may become profitable with technological advances or price increases; mineralized material too low in concentration to constitute ore, but from which ore may be formed through secondary enrichment.

[65] Technologically Enhanced Naturally Occurring Radioactive Materials from Uranium Mining; Volume 1: Mining and Reclamation Background; EPA 402-R-08-005; Revision April 2008. See http://www.epa.gov/radiation/tenorm/pubs.html#402-r-08-005.

46

U.S. Fourth National Report-Joint Convention on the Safety of Spent Fuel Management and on the Safety of Radioactive Waste Management

Table D-5 Summary of Decommissioning Activities in Progress		
Sector	Type	Number
Government	DOE Nuclear/Radioactive Facilities for which Decommissioning is Ongoing or Pending	1063
Government/ Commercial	Formerly Utilized Sites Remedial Action Program Sites (FUSRAP)	24[66]
	Decommissioning Materials Sites by Regulated NRC	20
	Decommissioning Material Sites in NRC Agreement States	44
Commercial	Nuclear Power Plants[67]	12
	Other Non-Power Reactor Facilities	12
	Uranium Recovery Facilities (NRC)	11
	Uranium Recovery Facilities (Agreement States)	14

D.3.1 DOE Sites with Decommissioning/Remediation Projects

The U.S. has a legacy of radioactive waste from past government activities spanning five decades. A total of 108 sites covering more than 800000 hectares (two million acres) of land have been used by the U.S. Government for nuclear research and development and nuclear weapons production activities. Most of the land at these sites is not contaminated. Within the boundaries of these sites are numerous radiological-controlled areas with thousands of individual facilities, encompassing 10500 discrete contaminated locations ("release sites"). Over 7000 of these release sites have been cleaned up (500 since the last report). Full remediation is complete at 89 of 108 DOE sites (4 since the last report), and 468 nuclear or radiological facilities are decommissioned (33 since the last report).

The U.S. Government continues to safeguard its nuclear materials, dispose of waste, remediate extensive surface and ground water contamination, and deactivate and decommission thousands of excess contaminated facilities. Annex D-4 shows a summary of the remaining nuclear/radioactive facility decommissioning projects and a summary of remaining DOE remediation projects.

D.3.2 Formerly Utilized Sites Remedial Action Program

The Formerly Utilized Sites Remedial Action Program (FUSRAP) began in 1974 to identify, investigate, and clean up or control sites where the Manhattan Engineer District and later the Atomic Energy Commission conducted defense and energy research activities in the 1940s - 1960s. Congress transferred FUSRAP management to the U.S. Army Corps of Engineers (USACE) in 1997. USACE continues to clean up sites started by or newly identified by DOE, and assigned by Congress. FUSRAP sites are distinct from the formerly licensed facilities, addressed in Section H.1.2.[68]

---

[66] Source: US Army Corps of Engineers, *Formerly Utilized Sites Remedial Action Program Update*, October 2010.
[67] Zion 1 & 2 counted as one facility.
[68] More FUSRAP information can be found at: https://environment.usace.army.mil/what_we_do/fusrap

FUSRAP sites are returned to DOE for long-term stewardship when remediation is completed. DOE's Office of Legacy Management manages sites and activities where missions have been completed and sites closed. This Office has responsibility for all FUSRAP sites remediated by DOE and those transferred back to DOE by USACE.[69] The contaminants at FUSRAP sites are primarily low levels of uranium, thorium, and radium, with their associated decay products. Materials containing low levels of radioactive residues are excavated, packaged, and transported for disposal at licensed commercial disposal sites, or to hazardous waste landfills, as appropriate. Annex D-5 lists FUSRAP sites with ongoing remediation. In some cases, the FUSRAP sites are also considered as complex material decommissioning sites, and are listed in Annex D-6.

D.3.3 Complex Materials Sites Decommissioning (NRC)

NRC regards complex sites as those that are required to provide a decommissioning plan or sites that require formal NRC or State approval prior to being decommissioned. NRC has taken a comprehensive approach to its decommissioning program to achieve better effectiveness. Of the 64 complex materials sites that are currently undergoing decommissioning in the U.S., NRC currently regulates 20 sites, located in 11 States, overseen by the agency's four regional offices. See Section F.6.1 for additional information. This comprehensive decommissioning program uses a dose-based approach for regulating decommissioning activities, and includes routine decommissioning sites, formerly licensed sites, non-routine/complex sites, fuel cycle sites, and test/research and power reactors. Remediating these sites is now managed more effectively as part of this larger program.

As of February 2011, 12 nuclear power and early demonstration reactors, 12 research and test reactors, 20 complex decommissioning materials facilities, 1 fuel cycle facility, 21 Title I uranium milling sites, and 11 Title II uranium recovery facilities are undergoing non-routine decommissioning or are in long-term safe storage, under NRC jurisdiction. Annex D-6 provides a list of these 20 "complex sites" subject to decommissioning. More specific information on the decommissioning status of NRC regulated sites can be found at NRC's website,[70] including specific status information for each complex site.

D.3.3.1 Complex Decommissioning Sites Regulated by NRC Agreement States

Under the provisions of the AEA, NRC can relinquish regulatory authority to individual states (Agreement State), including regulation of decommissioning material sites in those states. A recent example is the regulatory authority for decommissioning complex material sites in the Commonwealth of Virginia. Currently, 10 of the 37 Agreement States are regulating the decommissioning of 44 complex materials sites, with technical support from NRC's regional offices, as needed. See Section E.2.4.2 for additional information. Annex D-6 also lists those facilities undergoing decommissioning in the Agreement States.

---

[69] Extensive FUSRAP-related information (including information on specific sites) is available on the Legacy Management web page at http://www.lm.doe.gov. Legacy Management has also developed the Considered Sites Database (CSD) to provide public information documenting site eligibility and characterization, remediation, verification, and certification for all FUSRAP sites. CSD is available at http://www.lm.doe.gov/Considered_Sites/.

[70] See http://www.nrc.gov/about-nrc/regulatory/decommissioning.html.

D.3.3.2 Power and Non-Power Reactor Decommissioning

NRC has regulatory oversight responsibility for decommissioning 12 power reactors as of February 2011. NRC also provides oversight for decommissioning of research and test reactors. The 12 research and test reactors identified in Annex D-7 are undergoing decommissioning. In 2010, fuel was transferred from the General Atomics TRIGA Mark F and Mark I research reactors to DOE for storage. General Atomics is now able to proceed with decommissioning of these reactors.

D.3.3.3 Other Non-Power Facility Decommissioning

NRC provides project management and technical review for decommissioning and reclamation of facilities regulated in 10 CFR Part 40, Appendix A [under the Uranium Mill Tailings Radiation Control Act Title II]. These licensees include conventional uranium mills, heap leach facilities, and ISR facilities. Annex D-3 shows these sites. Decommissioning activities at conventional uranium mills include mill demolition, groundwater cleanup, soil cleanup, and closure of tailings impoundment. Decommissioning activities at ISR facilities are focused on restoring groundwater quality to pre-operational conditions, soil cleanup, and building demolition.

NRC also provides licensing oversight and decommissioning project management for fuel cycle facilities, including conversion plants, enrichment plants, and fuel manufacturing plants. NRC continues to work closely with the states and EPA to regulate remediation of unused portions of fuel cycle facilities. The only fuel cycle facility undergoing partial decommissioning is the Nuclear Fuels Services site in Erwin, Tennessee.

D.3.4 EPA Site Remediation

EPA remediates radiologically contaminated sites using its Comprehensive Environmental Response, Compensation, and Liability Act (CERCLA) authority. Since the passage of CERCLA in 1980, 56 radiologically contaminated sites have been placed on the National Priorities List (NPL) for CERCLA (out of 1,637 sites listed – 1,290 sites are currently on the NPL). Cleanup has been completed or the selected remedy implemented (e.g., construction of a groundwater treatment system that may operate over a number of years) at 37 of the radiologically contaminated sites. NPL sites have included uranium mines, DOE facilities, NRC licensees, and sites being addressed through FUSRAP.

49

U.S. Fourth National Report-Joint Convention on the Safety of Spent Fuel Management and on the Safety of Radioactive Waste Management

E. LEGISLATIVE & REGULATORY SYSTEMS

E.1 Legislative System

The policy on regulatory control of radioactive waste management in the United States of America (U.S.) has evolved through a series of laws establishing Federal Government agencies responsible for the safety of radioactive materials. Federal legislation is enacted by Congress and signed into law by the President. U.S. Laws apply to all 50 states and its territories. Legislation on the safety of spent fuel and radioactive waste can be traced back for five decades. Table E-1 identifies key U.S. Laws governing radioactive waste management and is supplemented by additional historical information in previous National Reports.

Table E-1 Key U.S. Laws Governing Radioactive Waste Management

Atomic Energy Act of 1954, as amended (AEA), established the Atomic Energy Commission (AEC), the predecessor to the Nuclear Regulatory Commission (NRC) and the Department of Energy (DOE), with Federal responsibility to regulate the use of nuclear materials including the regulation of civilian nuclear reactors. Under Reorganization Plan No. 3 of 1970, which created the U.S. Environmental Protection Agency (EPA), authority to establish generally applicable environmental standards was transferred to EPA along with authority to provide Federal guidance on radiation protection matters affecting public health.

The Price-Anderson Act (1957) was enacted to encourage development of the nuclear industry and ensure prompt and equitable compensation in the event of a nuclear incident. The Act provides a system of financial protection for persons who may be liable for and persons who may be injured by such an incident.

Solid Waste Disposal Act (SWDA) of 1965, as amended, requires environmentally sound methods for disposal of household, municipal, commercial, and industrial waste. The Resource Conservation and Recovery Act (RCRA) is an amendment to the SWDA.

National Environmental Policy Act (NEPA) of 1969, as amended, requires Federal agencies to consider environmental values and factors in agency planning and decision-making.

Clean Air Act (CAA) of 1970 is the comprehensive federal law that regulates air emissions from stationary and mobile sources.

The Marine Protection, Research, and Sanctuaries Act of 1972, also known as the Ocean Dumping Act, prohibits the dumping of material into the ocean unreasonably degrading or endangering human health or the marine environment.

Safe Drinking Water Act (SDWA) of 1972, as amended, protects public health by regulating the nation's public drinking water supply; it requires actions to protect drinking water and its sources: rivers, lakes, reservoirs, springs, and ground water wells.

Energy Reorganization Act of 1974, as amended, abolished the AEC and established NRC and the Energy Research and Development Administration (ERDA) — the predecessor of DOE.

Resource Conservation and Recovery Act of 1976, as amended, regulates the handling and disposal of hazardous wastes, which are generated mainly by industry, also requires that open dumping of all solid wastes be brought to an end throughout the country by 1983.

Department of Energy Organization Act (1977) brought together most of the Government's energy programs, as well as defense responsibilities that included the design, construction, and testing of nuclear weapons into the new Department of Energy. DOE was established on October 1, 1977, assuming the responsibilities of the Federal Energy Administration, ERDA, the Federal Power

---

Table E-1 Key U.S. Laws Governing Radioactive Waste Management

Commission, and parts and programs of several other Federal agencies.

Uranium Mill Tailings and Radiation Control Act (UMTRCA) of 1978, as amended, vested EPA with overall responsibility for establishing health and environmental cleanup standards for uranium milling sites and contaminated vicinity properties, NRC with responsibility for licensing and regulating uranium production and related activities, including decommissioning, and DOE with responsibility for remediating inactive milling sites and long-term monitoring of the decommissioned sites.

Comprehensive Environmental Response, Compensation, and Liability Act of 1980 (CERCLA) as amended, also known as Superfund, provided EPA with authority to address abandoned hazardous waste sites and outlined the process to be followed in identifying and remediating sites, including determination of cleanup levels and pursuit of contribution to the cleanup or cost recovery against parties deemed to have contributed to the contamination. CERCLA includes radionuclides as a hazardous substance.

Low-Level Radioactive Waste Policy Act of 1980 and Low-Level Radioactive Waste Policy Amendments Act of 1985 (LLRWPAA) gave the states – rather than the Federal Government – responsibility to provide disposal capacity for commercial Class A, B and C low-level waste (LLW); authorized the formation of regional compacts (groups of states) for the safe disposal of such LLW; and allowed compacts to decide whether to exclude waste generated outside the compact. The LLRWPAA gave the Federal government responsibility for the disposal of GTCC LLW that results from NRC or Agreement State licensed activities.

National Security and Military Applications of Nuclear Energy Authorization Act of 1980 Section 213 (a) of the Act authorizes Waste Isolation Pilot Plant (WIPP) "for the express purpose of providing a research and development facility to demonstrate the safe disposal of radioactive wastes resulting from defense activities and programs of the U.S. exempted from regulation by the U.S. Nuclear Regulatory Commission."

West Valley Demonstration Project Act of 1980 authorized DOE to conduct a technology demonstration project for solidifying high-level waste (HLW), disposing of waste created by the solidification, and decommissioning the facilities used in the process. The Act required DOE to enter into an agreement with the State of New York for carrying out the Project.

Nuclear Waste Policy Act of 1982 (NWPA) as amended by the Nuclear Waste Policy Amendments Act of 1987 (NWPAA) establishes the Federal responsibility for disposal of spent fuel and HLW.

Waste Isolation Pilot Plant Land Withdrawal Act (WIPP LWA) of 1992, as amended, withdraws land from the public domain for operation of the WIPP. Defines operational limitations and the role of the U.S. Environmental Protection Agency and the U.S. Mine Safety and Health Administration. Exempts transuranic (TRU) mixed waste destined for disposal at WIPP from treatment requirements and land disposal prohibitions under the Solid Waste Disposal Act. The Act provides for EPA continuing oversight role at WIPP, including recertification that WIPP meets EPA standards.

Energy Policy Act of 1992 (EnPA) mandated site-specific public health and safety standards and site-specific licensing requirements for the proposed repository at Yucca Mountain, Nevada. Among other things, it also authorized DOE to reimburse certain "active" uranium and thorium milling owners for a portion of their remedial action costs.

Energy Policy Act of 2005 (EPAct05) Sets forth an energy and development program and includes specific provisions addressing, among other things, disposal of greater-than-Class C Low-Level Waste (GTCC LLW) (including certain sealed sources), naturally occurring radioactive materials (NORM), and accelerator-produced waste.

52

U.S. Fourth National Report-Joint Convention on the Safety of Spent Fuel Management and on the Safety of Radioactive Waste Management

E.2　Regulatory System

The regulatory system for spent fuel and radioactive waste management in the U.S. involves several agencies: NRC, regulating the commercial nuclear sector; EPA, establishing environmental standards; and DOE, regulating its government programs. Some NRC regulatory authority — excluding spent fuel, special nuclear material sufficient to form a critical mass, and HLW — can be relinquished to the 50 states of the U.S. (including territories, Puerto Rico, and the District of Columbia) under its Agreement State Program. This is a provision of Section 274 of the Atomic Energy Act of 1954, as amended. This authority includes regulating commercial LLW disposal sites and uranium mill tailings sites, and regulatory authority over disposal of mill tailings. Some states also have regulatory authority delegated to them by EPA, such as for discharges from some industrial or mining practices. These are referred to as EPA Authorized States. See Section E.2.4.1.

The general regulations for the three Federal Agencies responsible for radioactive waste regulation are contained in Title 10 (for NRC and DOE) and Title 40 (for EPA) of the U.S. Code of Federal Regulations (CFR). U.S. Government regulations are developed through an open process, including the opportunity for public comment. New regulations are published in the Federal Register, in proposed or final forms. Specific regulations for each Agency are in Table E-2. Copies of these regulations are available in print and electronically

DOE Orders are internal directives, which function similar to regulations for DOE and DOE contractor activities. Compliance with orders is mandatory for DOE and is enforced through contract provisions for DOE contractors.

The separation between EPA's standard-setting function and NRC's implementing function reflects a nearly 40-year old Congressional policy of centralizing environmental standard setting in a single agency. When EPA was established, it was given environmental authorities scattered among several older agencies, including NRC's predecessor, the AEC. There are advantages to having an agency both set and implement standards, and NRC does so in many subject areas, most especially reactor design and operation. Nonetheless, there are also advantages to having environmental standards set on a national basis by a single agency whose jurisdiction is wide enough to permit the agency to rank risks from many sources, including nuclear

53

U.S. Fourth National Report-Joint Convention on the Safety of Spent Fuel Management and on the Safety of Radioactive Waste Management

Table E-2 Spent Fuel and Radioactive Waste Management Regulations

U.S. Nuclear Regulatory Commission

- 10 CFR Part 20, *Standards for Protection Against Radiation*
- 10 CFR Part 30, *Rules of General Applicability to Domestic Licensing of Byproduct Material*
- 10 CFR Part 40, *Domestic Licensing of Source Material*
- 10 CFR Part 51, *Environmental Protection Regulations for Domestic Licensing and Related Regulatory Functions*
- 10 CFR Part 60, *Disposal of High-Level Radioactive Wastes in Geologic Repositories (for sites other than Yucca Mountain)*
- 10 CFR Part 61, *Licensing Requirements for Land Disposal of Radioactive Waste*
- 10 CFR Part 62, *Criteria and Procedures for Emergency Access to Non-Federal and Regional Low-Level Waste Disposal Facilities*
- 10 CFR Part 63, *Disposal of High-level Radioactive Wastes in a Geologic Repository at Yucca Mountain, Nevada*
- 10 CFR Part 70, *Domestic Licensing of Special Nuclear Material*
- 10 CFR Part 71, *Packaging and Transportation of Radioactive Material*
- 10 CFR Part 72, *Licensing Requirements for the Independent Storage of Spent Nuclear Fuel, Reactor-Related Greater than Class C Waste, and High-Level Radioactive Waste*
- 10 CFR Part 110, *Export and Import of Nuclear Equipment and Material*

U.S. Department of Energy

- 10 CFR Part 765, *Reimbursement of Costs for Remedial Action at Active Uranium and Thorium Processing Sites*
- 10 CFR Part 766, *Uranium Enrichment Decontamination and Decommissioning Fund; Procedures for Special Assessment of Domestic Utilities*
- 10 CFR Part 820, *Procedural Rules for DOE Nuclear Facilities*
- 10 CFR Part 830, *Nuclear Safety Management*
- 10 CFR Part 835, *Occupational Radiation Protection*
- 10 CFR Part 960, *General Guidelines for the Recommendation for Sites for Nuclear Waste Repositories*
- 10 CFR Part 963, *Yucca Mountain Site Suitability Guidelines*
- 10 CFR Part 1021, *National Environmental Policy Act Implementing Procedures*

The following DOE directives are applicable to safety[71]:

- Order 151.1C, *Comprehensive Emergency Management System*
- Policy 226.1B, *Department of Energy Oversight Policy*
- Order 226.1B, *Implementation of Department of Energy Oversight Policy*
- Order 231.1A, *Environment, Safety, and Health*
- Order 360.1B, *Federal Employee Training*
- Order 414.1D, *Quality Assurance*
- Order 420.1B, *Facility Safety*
- Guide 421.1-2, *Guide 423.1-1; DOE Guide 424.1-1B, Implementation Guides for 10 CFR 830*
- Order 422.1, *Conduct of Operations*
- Order 425.1D, *Startup and restart of Nuclear Facilities*
- Order 430.1B, *Real Property Asset Management*
- Order 433.1B, *Maintenance Management Program*
- Order 435.1, *Radioactive Waste Management*
- Order 440.1B, *Worker Protection Management for DOE Federal and Contractor Employees*
- Order 470.2B, *Independent Oversight and Performance Assurance Program*
- Order 458.1, *Radiation Protection of the Public and the Environment*
- Order 462.2, *Personnel Selection, Qualification, and Training Requirements for DOE Nuclear Facilities*

---

[71] For DOE Directives see: https://www.directives.doe.gov/.

54

U.S. Fourth National Report-Joint Convention on the Safety of Spent Fuel Management and on the Safety of Radioactive Waste Management

Table E-2 Spent Fuel and Radioactive Waste Management Regulations
U.S. Environmental Protection Agency

40 CFR Part 61, National Emission Standards for Hazardous Air Pollutants
- Subpart B, *Radon from Underground Uranium Mines*
- Subpart H, *Radionuclide Emissions, other than Radon, from DOE Facilities*
- Subpart I, *Radionuclide Emissions from Federal Facilities other than DOE or NRC Licensed Facilities*
- Subpart K, *Radionuclide Emissions from Elemental Phosphorus Plants*
- Subpart Q, *Radon from DOE Facilities*
- Subpart R, *Radon from Phosphogypsum Stacks*
- Subpart T, *Radon from Disposal of Mill Tailings*
- Subpart W, *Radon from Tailings at Operating Mills*

40 CFR Part 190, *Environmental Radiation Protection Standards for Nuclear Power Operations*
40 CFR Part 191, *Environmental Radiation Protection Standards for Management and Disposal of Spent Nuclear Fuel, High-level and Transuranic Radioactive Wastes*

40 CFR Part 192, *Health and Environmental Protection Standards for Uranium and Thorium Mill Tailings*

40 CFR Part 194, *Criteria for the Certification and Recertification of the Waste Isolation Pilot Plant's (WIPP) Compliance with the 40 CFR Part 191 Disposal Regulations*

40 CFR Part 197, *Public Health and Environmental Radiation Protection Standards for Yucca Mountain, Nevada*

Other Title 40, Code of Federal Regulations relating to radiation protection include:
- Part 141, *National Primary Drinking Water Regulations*
- Part 147, *State Underground Injection Control Programs*
- Part 148, *Hazardous Waste Injection Restrictions*
- Part 195, *Radon Proficiency Programs*
- Parts 220 and 133, *Ocean Dumping*
- Part 300, *National Oil and Hazardous Substances Pollution Contingency Plan*
- Part 302, *Designation, Reportable Quantities, and Notification*
- Part 440, *Ore Mining and Dressing Point Source Category (Uranium, Radium, and Vanadium Ores subcategory)*

E.2.1 U.S. Nuclear Regulatory Commission

NRC is an independent regulatory agency created from the former AEC by Congress under the Energy Reorganization Act of 1974 to assure protection of the public health and safety and the environment, and to promote the common defense and security in the civilian use of byproduct, source, and special nuclear materials.

NRC regulates:
- Commercial nuclear power, non-power research, test, and training reactors;
- Fuel cycle facilities, medical, academic, and industrial uses of nuclear materials;
- Storage and disposal of nuclear materials and waste; and
- Certain DOE activities and facilities over which Congress has provided NRC licensing and related regulatory authority.

NRC regulates manufacture, production, transfer or delivery, receiving, acquisition, ownership, possession, and use of commercial radioactive materials, including the regulation of the associated radioactive waste. The key elements of NRC's regulatory program are described in detail at: http://www.nrc.gov; this information is also available from previous National Reports.

Specifically, NRC regulates management and disposal of LLW and HLW and decontaminating and decommissioning facilities and sites. NRC is also responsible for establishing the technical bases for regulations and provides information and technical bases for developing acceptance criteria for licensing reviews. A listing of guidance issued by NRC is provided in Annex E-1.

An important aspect of NRC's regulatory program is inspection and enforcement. NRC has four regional offices, which inspect licensed facilities in their regions, including nuclear waste facilities.[72] NRC's Office of Federal and State Materials and Environmental Management Programs communicates with state and local governments, and Tribal Governments, and oversees the Agreement State Program. NRC Agreement States are discussed in Section E.2.4.2.

E.2.1.1 Uranium Recovery Regulation

NRC is responsible for planning and implementing regulatory programs under UMTRCA. UMTRCA amended the AEA to require EPA to issue generally applicable standards for controlling uranium mill tailings. EPA issued standards for both Title I (residual radioactive material) and Title II (AEA Section 11e.(2) byproduct material) sites in 1983. The Title I program established a joint Federal/state funded program for remedial action at abandoned mill tailings sites, with final Federal ownership under NRC license (see Section D.2.2.3.1). NRC, under Title I, must evaluate DOE designs and agree DOE actions meet standards set by EPA. NRC and DOE have a memorandum of understanding to clarify their roles and responsibilities, e.g., to minimize or eliminate duplication of effort between the two agencies.

UMTRCA Title II involves planning and directing activities for active, licensed uranium recovery facilities, including facility licensing and operation, and mill tailings management and decommissioning. Title II deals with NRC or Agreement States sites. NRC regulations are found in 10 CFR Part 40, Appendix A, and are consistent with EPA Title II standards and meet UMTRCA requirements. NRC has authority under Title II to control radiological and non-radiological hazards, and ensure NRC-licensed and Agreement State-licensed sites meet all standards and requirements during operations and before termination of licenses. NRC reviews Title II license applicant's plans for operating, reclaiming, decommissioning, and groundwater corrective action; license applications and renewals; license conditions changes; and, annual surety updates. Long-term care provisions are addressed in Section D.2.2.3.

NRC also provides technical assistance to Agreement States on uranium recovery issues and implements an active interface program including consultation with Federal agencies, states, Tribal governments, and other entities to promote understanding of uranium programs and resolving concerns in a timely manner.

---

[72] Specific information on NRC Regional Offices can be accessed at: http://www.nrc.gov/about-nrc/organization.html.

E.2.1.2 HLW and Spent Fuel Regulation

Regulatory responsibility for disposal of HLW and spent fuel is described in the Energy Reorganization Act, the Nuclear Waste Policy Act of 1982, as amended (NWPA), and Energy Policy Act of 1992. NRC has licensing authority for facilities for the disposal of spent fuel or HLW, including:

- Conducting pre-licensing consultation;
- Certifying transportation packages;
- Hosting meetings at NRC Headquarters and in Nevada and other affected states;
- Implementing and maintaining the HLW Licensing Support Network; and
- Review of License Application for construction authorization of a repository at Yucca Mountain.

EPA has issued final standards for HLW disposal at Yucca Mountain (at 40 CFR Part 197), and NRC has published conforming licensing regulations for HLW disposal at Yucca Mountain (at 10 CFR Part 63). EPA has also issued final standards for HLW disposal for sites other than Yucca Mountain (at 40 CFR Part 191). Although NRC finalized its regulations for HLW disposal for sites other than Yucca Mountain, these regulations have remained substantially the same since its development approximately 30 years ago. NRC is currently evaluating the need for revisions to these regulations to conform to EPA's final standards and to take advantage of regulatory enhancements that have occurred since 10 CFR Part 60 was developed.

EPA standards and NRC regulations are generally consistent with national and international recommendations for radiation protection standards.

E.2.1.3 LLW Regulation

Commercial LLW disposal facilities are designed, constructed, and operated under licenses issued by either NRC or an Agreement State, based on NRC health and safety regulations governing waste disposal quantities, forms, and activity levels (10 CFR Part 61, *Licensing Requirements for Land Disposal of Radioactive Waste*). This regulation establishes the procedures, criteria, and terms and conditions for the issuance of licenses for the disposal of LLW. Four performance objectives, including protection of an inadvertent intruder into the waste disposal site, define the overall level of safety to be achieved by disposal.[73] Specifically, Section 61.55 addresses the classes of LLW. These classes are described in Table B-1 of this report. The process by which a state becomes an Agreement State regulating LLW is described in http://www.nrc.gov/about-nrc/state-tribal/become-agreement.html#sa700.

The LLRWPAA gave the states responsibility for the disposal of low-level radioactive waste generated within their borders (except for certain waste generated by the Federal government). The Act authorized the states to enter into compacts that would allow them to dispose of waste at a common disposal facility. States were in various stages of planning, siting and licensing LLW disposal facilities in the late 1980s and early 1990s in an attempt to meet the milestones of LLRWPAA. NRC developed a *Standard Format*

---

[73] The other performance objectives are protection of the general population from releases of radioactivity; protection of individuals during the operation of the facility (as opposed to after the facility is closed) and stability of the disposal site.

and *Content* guide (NUREG-1199) and a *Standard Review Plan* (NUREG-1200), providing guidance on licensing LLW disposal facilities enabling NRC to meet its statutory requirements of reviewing a license application within 15 months of receipt and to provide technical guidance to Agreement States. NRC published a final report, *A Performance Assessment Methodology for Low-Level Radioactive Waste Disposal Facilities: Recommendations of NRC's Performance Assessment Working Group* (NUREG-1573) in October 2002. NRC published[74] the results of the staff's strategic assessment of NRC's LLW regulatory program in October 2007. The results include a prioritized listing of ongoing and future staff actions and activities, along with associated schedules and resource estimates. These included activities that would best allow NRC to accommodate changing priorities and disposal challenges within the States. See Section H.1.1. for additional information on commercially licensed LLW disposal facilities and disposal compacts.

E.2.1.4 Decommissioning Regulation

Decommissioning involves safely removing a facility from service and reducing residual radioactivity to a level permitting the property to be released for unrestricted or restricted use. This action is taken by a licensee before NRC terminates its license.

Title 10 CFR Part 20, Subpart E provides the main decommissioning requirements and regulations; however, the diversity and complexity of decommissioning dictates regulations focusing on specific features associated with the different types of sites and activities undergoing termination of licenses and decommissioning.[75] A unique consideration for decommissioning in the U.S. is a timeliness provision; i.e., specific time periods for decommissioning unused portions of operating nuclear materials facilities and for decommissioning the entire site upon termination of operations.

NRC has developed a number of guidance documents to help licensees prepare decommissioning documents (see Annex E-1). The primary decommissioning guidance is documented in the *Consolidated Decommissioning Guidance* (NUREG-1757) and the *Standard Review Plan for Evaluating Nuclear Power Reactor License Termination Plans* (NUREG-1700, Rev. 1). These documents describe: (1) acceptable methods for implementing NRC's regulations; (2) techniques and criteria used by NRC in evaluating decommissioning actions; and (3) guidance to licensees responsible for decommissioning NRC-licensed sites.[76] he *Consolidated Decommissioning Guidance* documents are periodically revised and re-issued as a result of lessons learned changes in regulations and an ongoing continuous improvement activity. For additional information on these activities, please refer to past issues of the U.S. National Report and the Status of the Decommissioning Program: 2010 Annual Report, which is accessible at: http://www.nrc.gov/about-nrc/regulatory/decommissioning.html.

---

[74] Strategic Assessment of Low-Level Radioactive Waste Regulatory Program, available at http://pbadupws.nrc.gov/docs/ML0713/ML071350350.pdf.

[75] The specific regulations are accessible at http://www.nrc.gov/about-nrc/regulatory/decommissioning/reg-guides-comm/regulations.html.

[76] These are accessible from http://www.nrc.gov/about-nrc/regulatory/decommissioning/reg-guides-comm/guidance.html.

58

U.S. Fourth National Report-Joint Convention on the Safety of Spent Fuel Management and on the Safety of Radioactive Waste Management

Additional information on NRC's decommissioning approach is provided in Section F.6.1 of this report. There are many opportunities for public involvement and information throughout the decommissioning process.[77,78]

The decommissioning process for non-power reactor facilities can be initiated by any number of conditions.[79] These include expiration of the license and cessation of operations in all or part of the site for 24 months. NRC inspects the licensee's decommissioning operations to ensure compliance with the decommissioning plan. These inspections will normally include in process and confirmatory radiological surveys.[80]

In the final steps of decommissioning of material sites, licensees are required to perform a number of actions including certification of the disposition of all licensed material and performance of a radiation survey of the premises.[81]

E.2.1.5 NRC's Integrated Materials Performance Evaluation Program

NRC, in coordination with the Agreement States, developed and piloted a review process in 1994 for Agreement State and NRC Regional materials programs called the Integrated Materials Performance Evaluation Program (IMPEP). Common performance indicators were established to obtain comparable information on the performance of each program. NRC began full implementation of IMPEP in 1996 to ensure public health and safety are adequately protected from potential hazards of using radioactive materials, and Agreement State programs are compatible with NRC's program.[82]

IMPEP employs a team of NRC and Agreement State staff to assess both Agreement State and NRC radioactive materials licensing and inspection programs. All reviews use the following common indicators in the assessment and place primary emphasis on performance:

- Technical Staffing and Training;
- Status of Materials Inspection Program;
- Technical Quality of Inspections;
- Technical Quality of Licensing Actions; and
- Technical Quality of Incident and Allegation Activities.

---

[77] See http://www.nrc.gov/about-nrc/regulatory/decommissioning/public-involve.html.

[78] Additional information on the decommissioning process for reactors is accessible at: http://www.nrc.gov/about-nrc/regulatory/decommissioning/process.html.

[79] Major steps in the complex materials site decommissioning process are described at http://www.nrc.gov/about-nrc/regulatory/decommissioning/process.html.

[80] See NRC's decommissioning oversight activities at http://www.nrc.gov/about-nrc/regulatory/decommissioning/oversight.html.

[81] Specific details for unrestricted versus restricted release, schedules for notification and completion of decommissioning milestones, as well as alternatives in the compliance with regulatory requirements for decommissioning are discussed in greater detail at http://www.nrc.gov/about-nrc/regulatory/decommissioning.html.

[82] The IMPEP program was selected in 2004 as among the top 50 programs for the "Innovations in American Government Awards," sponsored by the Ash Institute for Democratic Governance and Innovation at Harvard University's John F. Kennedy School of Government and administered in partnership with the Council for Excellence in Government.

59

Additional areas are identified as non-common performance indicators (Compatibility Requirements, Sealed Source and Device Evaluation Program, Low-Level Radioactive Waste Disposal Program, and Uranium Recovery Program) and may also be addressed in the assessment.

Both Agreement States and NRC Regional and Headquarters Offices are reviewed under this program. About 10-12 reviews are scheduled each year. Regions and Agreement States are routinely reviewed every four years, although the frequency may be decreased based on good performance or increased based on poor performance. The final determination of adequacy of each NRC Regional program and both adequacy and compatibility of each Agreement State program, is made by a Management Review Board (MRB), using the review team's report as a basis for its determination. This Board is composed of NRC managers and an Agreement State program manager who serves as an Agreement State liaison to the MRB.

The Organization of Agreement States is invited to nominate liaisons to participate in MRB meetings as a nonvoting participant. State representatives receive all relevant documents and engage in all MRB discussions except those potentially involving the Agreement State liaison's own state. Agreement States and Regional representatives are also invited to attend their individual MRB meetings to discuss IMPEP team findings with the MRB.

The range of possible findings for an Agreement State program are:

1. Adequate to protect public health and safety and compatible/not compatible;
2. Adequate, but needs improvement and compatible/not compatible; and
3. Inadequate to protect public health and safety and compatible/not compatible.

NRC Offices are rated in the same manner, but without the additional compatibility finding. IMPEP good practices and lessons learned are made available to all regulatory programs.[83] Additional information on the IMPEP program can be found at the IMPEP Toolbox. Lessons learned reflect input and feedback from Agreement State officials and NRC regional staff.

E.2.2 U.S. Environmental Protection Agency

EPA has several radioactive waste regulatory functions. These areas are described in more detail below.

E.2.2.1 Waste Isolation Pilot Plant Oversight

EPA issues radiation standards and certifies compliance of the WIPP disposal facility. The Waste Isolation Pilot Plant Land Withdrawal Act (WIPP LWA), as amended, required EPA to issue final regulations for disposal of spent fuel, HLW, and TRU waste. It also gave EPA authority to develop criteria implementing final WIPP radioactive waste disposal standards. EPA must also determine every five years whether or not the WIPP facility is in compliance with applicable standards. The WIPP LWA also requires EPA to determine whether WIPP complies with other Federal environmental and public health and safety regulations, such as the Clean Air Act and the Solid Waste Disposal Act.

---

[83] Additional information on IMPEP may be found at http://nrc-stp.ornl.gov/impeptools.html.

60

U.S. Fourth National Report-Joint Convention on the Safety of Spent Fuel Management and on the Safety of Radioactive Waste Management

EPA issued final amendments to its radioactive waste disposal standards for spent fuel, HLW, and TRU radioactive waste on December 20, 1993, initially promulgated in 1985 (40 CFR Part 191). The final individual protection standards require disposal systems to limit the amount of radiation to which an individual can be exposed for 10,000 years. The final groundwater protection standards require disposal systems to be designed so that for 10000 years after waste disposal, contamination in off-site underground sources of drinking water will not exceed the maximum contaminant level for radionuclides established by EPA under the Safe Drinking Water Act. Containment requirements of Subpart C limit releases of radionuclides to specified levels for 10000 years after the facility accepts its final waste for disposal, while assurance requirements involve additional measures intended to provide confidence in the long-term containment of radioactive waste. EPA issued final compliance criteria on February 9, 1996 (40 CFR Part 194) for certification and recertification of WIPP compliance with the final radioactive waste disposal standards in 40 CFR Part 191.

DOE submitted a Compliance Certification Application to EPA on October 29, 1996, to demonstrate WIPP complies with the criteria at 40 CFR Part 194. EPA then conducted a very open certification review process, involving multiple opportunities for written public comments and public hearings. EPA issued a Final Rulemaking Notice on the certification decision on May 18, 1998. WIPP received its first TRU waste shipment on March 26, 1999.

DOE submitted an application for WIPP recertification in March 2004, as required by statute every five years. EPA issued its decision to recertify DOE's compliance with the applicable standards on April 10, 2006. DOE submitted the second Compliance Recertification Application to EPA on March 24, 2009. A recertification decision from EPA was made on November 18, 2010.

The Office of Radiation and Indoor Air coordinates most EPA actions under the WIPP LWA. Other EPA offices also play important roles concerning WIPP. EPA's Region VI office, based in Dallas, Texas, is responsible for determining WIPP compliance with all applicable environmental laws and regulations other than the radioactive waste disposal standards. The Region VI office also coordinates with EPA's Office of Solid Waste on hazardous waste issues. Some TRU waste intended for disposal at the WIPP also contains hazardous components, subjecting it to the regulations developed under the Resource Conservation and Recovery Act of 1976 (RCRA), as amended.

EPA conducts inspections of both waste generators and WIPP operations. Separate inspections may be conducted for waste characterization activities, quality assurance, or WIPP site activities (procedural or technical).

The State of New Mexico is authorized by EPA to carry out the base RCRA and mixed waste programs in lieu of equivalent Federal programs. The New Mexico Environment Department reviews permit applications for treatment, storage, and disposal facilities for hazardous waste, under Subtitle C of RCRA. WIPP's Hazardous Waste Facility Permit (HWFP) is renewed every 10 years. The updated HWFP was approved on November 30, 2010, and the permit's effective date was December 30, 2010.

61

U.S. Fourth National Report-Joint Convention on the Safety of Spent Fuel Management and on the Safety of Radioactive Waste Management

E.2.2.2 EPA HLW and Spent Fuel Disposal Standards

In addition to EPA's generally applicable standards in 40 CFR Part 191, the Energy
Policy Act of 1992 (EPAct92) required EPA to develop radiation protection standards
specifically for the Yucca Mountain site to protect the public and the environment from
exposure to radioactive wastes disposed in the repository (40 CFR Part 197). This was
issued in 2001. As a result of a legal challenge to the regulatory compliance period,
EPA amended the standard to extend the compliance period and incorporate other
supporting provisions. The amended standards were issued in 2008. Further legal
challenge to the amended standards is stayed pending resolution of DOE's motion to
withdraw its license application.

E.2.2.3 Mixed Waste Regulation

A dual regulatory framework exists for mixed waste. EPA or authorized states regulate
the hazardous waste component and NRC, NRC Agreement States, or DOE regulate
the radioactive component. NRC and DOE regulate mixed waste radiation hazards
using Atomic Energy Act of 1954 (AEA) authority. EPA regulates mixed waste chemical
hazards under its Resource Conservation and Recovery Act (RCRA) authority. NRC is
authorized by the AEA to issue licenses to commercial users of radioactive materials.
RCRA gives EPA authority to control hazardous waste from "cradle-to-grave". Waste
handlers must comply with both AEA and RCRA statutes and regulations once a waste
is found to be a mixed waste. The requirements of RCRA and AEA are generally
consistent and compatible. The provisions in Section 1006(a) of RCRA allow the AEA to
take precedence if provisions of requirements of the two acts are inconsistent.

Land Disposal Restriction regulations, under the 1984 Amendments to RCRA, prohibit
disposal of most mixed waste until it meets specific treatment standards for hazardous
constituents, which may be based on a concentration or a specific treatment technology.
Most commercial mixed waste (generated and stored) can be treated to meet Land
Disposal Restriction regulations with commercially available treatment technology. No
treatment or disposal capacity is available for a small percentage of commercial mixed
waste. Commercial mixed waste volumes are very small (approximately two percent)
compared to the total volume of mixed waste being generated or stored by DOE.

DOE has developed Site Treatment Plans to handle its mixed wastes under the Federal
Facilities Compliance Act, signed into law on October 6, 1992. These plans are being
implemented by orders issued by EPA or the state regulatory authority.

EPA issued regulations in 2001 allowing mixed waste to be exempted from RCRA
hazardous waste requirements, as long as it meets NRC or Agreement State
requirements. These regulations may be found at 40 CFR Part 266, Subpart N, and
apply to:

- Storage at the generator site or another site operating under the same license;
- Treatment in a tank or container at the generator site or another site operating
  under the same license;
- Transportation to a licensed treatment facility or low-level waste disposal facility;
  and

- Disposal at a licensed low-level waste disposal facility, as long as the waste meets RCRA treatment standards for hazardous constituents.

E.2.2.4 Uranium Mining and Milling Standards

UMTRCA, which amended the AEA, directed EPA to establish standards for active and inactive uranium and thorium mill sites (see Section D.2.2.3). The standards for active sites were issued in 1983 as 40 CFR Part 192 (and amended in 1995), and establish limits on radon emanations from tailings as well as contamination limits for buildings, soil, and ground water. NRC incorporated these standards into its regulations in 40 CFR Part 40, Appendix A. A key aspect of UMTRCA is that it required EPA's standards to address non-radiological contaminants in a manner consistent with EPA's requirements for managing chemically hazardous waste. The inactive site standards, 40 CFR Part 191, are implemented by DOE at inactive sites.

The AEA does not identify uranium-mining overburden as radioactive material to be controlled, and NRC and DOE do not regulate the disposition of conventional mining wastes as part of the nuclear fuel cycle. Once uranium-mining product is beneficiated or is brought into the milling circuit, including production from in situ recovery operations, then NRC and its Agreement States regulate its possession, use, transport, etc.

EPA has also established National Emission Standards for Hazardous Air Pollutants (NESHAPs) under the Clean Air Act for airborne radionuclide emissions from a variety of industrial sources (40 CFR Part 61). Subparts B, T, and W apply to underground uranium mines, inactive uranium mill tailings piles, and active uranium mill tailings piles, respectively. See Section E.2.2.4 of the Third National Report for more information on these regulations.

EPA is currently reviewing 40 CFR Part 192 and Subpart W for the purpose of updating the standards to more directly address in situ recovery facilities and reflect new risk assessments.

E.2.2.5 Other EPA Radiation-Related Authorities

EPA has regulatory responsibilities for a variety of other man-made and naturally occurring radioactive wastes:

- Developing general radiation protection guidance to the Federal government. Section F contains additional information about radiation protection;
- Limiting airborne emissions of radionuclides. Subpart H of EPA's NESHAPs standards limit the airborne emissions of radionuclides (other than radon) from DOE sites managing defense-related spent fuel and radioactive waste. A limit of 0.1 mSv (10 mrem) per year effective dose equivalent is applied to any member of the public in the vicinity of such sites. Emission monitoring is specified, and DOE sites are required to submit an annual compliance report to EPA. Subpart I applied similar requirements to NRC-licensed facilities but was rescinded when NRC established comparable requirements;
- Setting drinking water regulations, under the Safe Drinking Water Act, as amended, including standards for radionuclides in community water systems;

- The Clean Water Act authorizes the National Pollutant Discharge Elimination System (NPDES) permit program, which controls water pollution by regulating point sources that discharge pollutants into surface waters. Point sources are discrete conveyances such as pipes or man-made ditches. These permits apply to such activities as dewatering mines to allow resource recovery of minerals. In most cases, the NPDES permit program is administered by EPA authorized states;

- Remediating radiologically contaminated sites listed on the CERCLA National Priorities List (NPL). See Section D.3.4. The NPL includes sites licensed by NRC or Agreement States, as well as DOE sites. EPA and NRC entered into a Memorandum of Understanding (MOU) in October 2002, to avoid future confusion about the potential for dual regulation at decommissioned sites. This MOU defines conditions where the two agencies would consult on the decommissioning of NRC-licensed facilities;[84]

- Coordinating with state radiation protection agencies to protect the environment, workers, and the public from naturally occurring radioactive materials exposed or concentrated by mining or processing; and

- Coordinating with DOE, NRC and states on orphaned sources, recycled materials, and controlling imports and exports to prevent radioactively contaminated scrap from entering the U.S.[85] See further discussions in Sections I and J.

EPA is composed of a headquarters organization and 10 regional offices. Each EPA Regional Office is responsible for executing Agency's programs with states in its region. EPA also has 17 laboratories located across the nation.

E.2.3 U.S. Department of Energy

DOE is responsible for regulating its spent fuel and radioactive waste management activities pursuant to the Atomic Energy Act (see Table E-1), except in cases where Congress has specifically provided NRC with licensing and related regulatory authority over DOE activities or facilities. Radiation and environmental protection are ensured by a rigorous framework of Federal regulations, DOE Orders and Directives, and external recommendations by the Defense Nuclear Facilities Safety Board.

The major applicable Federal regulations include 10 CFR Part 820, *Procedural Rules for DOE Nuclear Activities*,[86] 10 CFR Part 830, *Nuclear Safety Management*,[87] and 10 CFR Part 835, *Occupational Radiation Protection*.[88]

---

[84] See OSWER Directives 9295.8-06 and 9295.8-06a at
http://www.epa.gov/superfund/health/contaminants/radiation/mou.htm .
[85] The U.S. Coast Guard and the U.S. Department of Homeland Security Customs and Border Protection have the lead in detecting and taking steps to prevent the illegal entry of such materials. They have the authority to take enforcement actions and, depending on the circumstances, may seize or have a shipment returned to the point of origination.
[86] See: http://ecfr.gpoaccess.gov/cgi/t/text/text-idx?c=ecfr&sid=878197d37c47d5d267017ca6fde249ba&rgn=div5&view=text&node=10:4.0.2.5.24&idno=10.
[87] See: http://ecfr.gpoaccess.gov/cgi/t/text/text-idx?c=ecfr&tpl=/ecfrbrowse/Title10/10cfr830_main_02.tpl.
[88] See: http://ecfr.gpoaccess.gov/cgi/t/text/text-idx?c=ecfr&tpl=/ecfrbrowse/Title10/10cfr835_main_02.tpl.

DOE regulates facility operations and radiation protection through standards and requirements established in DOE Orders and Directives. The major applicable orders include DOE Order 458.1, *Radiation Protection of the Public and the Environment*[89] and DOE Order 435.1, *Radioactive Waste Management.*[90] Table E-2 provides a list of spent fuel and radioactive waste management Federal regulations, and DOE Orders and Directives.

DOE implements applicable radiation protection standards considering and adopting, as appropriate, recommendations of authoritative organizations such as the National Council on Radiation Protection and Measurements and the International Commission on Radiological Protection. It is also DOE policy to adopt and implement standards generally consistent with those of NRC.

The following sections describe the multiple independent oversight and regulatory layers governing DOE's spent fuel and radioactive waste management activities.

E.2.3.1 DOE Independent Oversight

DOE spent fuel and radioactive waste management activities designated under the Joint Convention receive comprehensive independent oversight from DOE's Office of Health, Safety and Security (DOE-HSS). Oversight activities are described in DOE Order 470.2B, *Independent Oversight and Performance Assurance Program,*[91] and DOE Order 426.1A, *Implementation of Department of Energy Oversight Policy,*[92] and include:

- Ensuring DOE activities conform with applicable laws and requirements for protecting the environment and health and safety of the public, and the workers at DOE facilities;
- Conducting scientific and technical programs to enhance DOE's ability to protect the health and safety of workers and the public;
- Developing effective, efficient, and state-of-the-art environmental, occupational safety and health, and medical policies and rules for operating DOE facilities;
- Providing technical assistance to DOE programs to identify and resolve environment, safety, health, safeguards, and security issues; and
- Ensuring compliance with nuclear safety requirements.

DOE-HSS under DOE Manual 435.1-1, *Radioactive Waste Management Manual*[93] (Chapter I), establishes independent oversight of radioactive waste management and decommissioning programs to determine compliance with DOE environment, safety and health requirements and applicable EPA and state regulations. DOE regulations in 10 CFR Parts 820, 830, and 835 make DOE nuclear safety requirements subject to enforcement, including the imposition of civil and criminal penalties.

DOE-HSS develops, manages, and directs programs for health and safety policy to protect health and safety of workers, and for facility and systems operations safety. It

---

[89] See: https://www.directives.doe.gov/directives/current-directives/458.1-BOrder-admc2/view
[90] See: https://www.directives.doe.gov/directives/current-directives/435.1-BOrder-c1/view.
[91] See: https://www.directives.doe.gov/directives/restrict/470.2-BOrder-b/view.
[92] See: https://www.directives.doe.gov/directives/current-directives/226.1-BOrder-b/view.
[93] See: https://www.directives.doe.gov/directives/current-directives/435.1-DManual-1c1/view.

65

U.S. Fourth National Report-Joint Convention on the Safety of Spent Fuel Management and on the Safety of Radioactive Waste Management

serves as the primary DOE liaison with the Department of Labor Occupational Safety and Health Administration and NRC on health and safety regulation reviews and pending regulatory reform. A more complete description of DOE-HSS oversight functions are in the Third U.S. National Report.

DOE's Office of Price-Anderson Enforcement maintains the internal self-regulatory program; investigates potential violations; and, where warranted, initiates enforcement action including recommending whether civil penalties should be imposed. Those actions are performed according to processes and procedures in 10 CFR Part 820.

DOE's Office of the General Counsel (and the National Nuclear Security Administration Office of the General Counsel, as appropriate) ensures programs and processes are conducted in accordance with applicable Federal statutes and regulations. In particular, it exercises approval authority for DOE NEPA analyses. It coordinates with and assists in preparing environmental impact statements for major DOE proposed actions. It also develops written orders, policies, regulations, and guidance documents for environmental review requirements and their implementation.

E.2.3.2 Defense Nuclear Facilities Safety Board

The Defense Nuclear Facilities Safety Board (DNFSB) is an independent Federal agency established by Congress in 1988. DNFSB's mandate under the AEA is to make nuclear safety recommendations concerning DOE defense nuclear facilities. DNFSB reviews and evaluates the content and implementation of DOE health and safety standards for design, construction, operation, and decommissioning of defense nuclear facilities. DNFSB must then recommend to the Secretary of Energy any specific measures, such as changes in content and implementation of those standards, DNFSB believes should be adopted to ensure the public health and safety are adequately protected. DNFSB also reviews the design of new defense nuclear facilities before construction begins, as well as modifications to older facilities, and is required to recommend changes to protect health and safety.

E.2.3.3 Other Federal Regulators

Certain DOE facilities and operations are subject to regulation and independent oversight by other agencies as described in Section E.2 above. Both NRC and EPA regulate facilities within the DOE complex; e.g., the Three Mile Island damaged fuel and core debris are in NRC-licensed dry storage at the DOE Idaho Site. EPA certifies the WIPP (Section D.2.2.1) through its WIPP LWA authority (Section E.2.2.1). A complete list of spent fuel and radioactive waste management facilities and their licensing authority are in Annex D-1 and Annex D-2.

E.2.3.4 State Authorities

EPA authorized states play a significant role in regulation and independent oversight of DOE facilities (Section E.2.4.1). Most of DOE's cleanup is performed under CERCLA through Federal Facility Agreements (FFAs), and under RCRA through various consent and compliance orders. These enforceable regulatory agreements and orders with Federal and state agencies establish the scope of work to be performed at a given site and the dates by which specific cleanup milestones must be achieved. Failure to comply with these agreements and orders is subject to fines and penalties. Table E-3 describes the types of regulatory agreements. The status of cleanup projects can be found at http://www.em.doe.gov/pdfs/NDAApercent20Report-(01-15-09)a.pdf.

DOE has successfully developed a close working relationship with Tribal governments, state regulators, and local citizens. DOE's Office of Environmental Management has the largest Federal Advisory Committee Act chartered citizen advisory board in the Federal Government with boards at eight cleanup sites. DOE also supports working groups with the National Governors' Association, the National Conference of State Legislators, the Energy Communities Alliance (represents local communities at cleanup sites), and the State and Tribal Government Working Group.

Table E-3 Regulatory and Agreement Order Descriptions	
Agreement/ Order	Description
Federal Facility Agreement (FFA)	A legal agreement between DOE, EPA, and sometimes the state. It sets forth schedules and processes for site cleanup under CERCLA, including enforcement provisions for non-compliance. FFAs that include the state as a party often incorporate RCRA compliance requirements, as well as state hazardous waste law requirements that flow from RCRA.
Consent Order Or Consent Agreement Or Settlement Agreement	A legal agreement between DOE and EPA or the state, documenting the settlement of a cleanup issue outside of court. Consent orders, consent agreements, and settlement agreements are legally binding, so compliance disputes may ultimately be taken to court. Most consent orders, consent agreements, and settlement agreements address RCRA issues or state hazardous waste issues that flow from RCRA, although they can also address CERCLA issues. A few also incorporate Toxic Substances Control Act requirements.
Consent Decree	A court-issued enforceable order, generally reflecting an agreement between DOE and EPA or the state. Consent decrees can cover CERCLA or RCRA, as well as state hazardous waste laws.
Site Treatment Plan and Compliance Order	A legal agreement and plan developed under the Federal Facility Compliance Act and RCRA for DOE facilities that generate or store mixed wastes, setting schedules to treat all of the facilities' mixed waste.

E.2.3.5 Nuclear Waste Technical Review Board

The U.S. Nuclear Waste Technical Review Board (NWTRB) was created by Congressional legislation, in the 1987 amendments to the Nuclear Waste Policy Act (NWPAA). It advises both Congress and the Secretary of Energy on technical issues related to DOE's implementation of the NWPA. The Board evaluates the technical validity of all activities undertaken by the Secretary of Energy related to DOE's obligation to manage and develop an approach to dispose of spent fuel and HLW. The NWTRB is a unique Federal agency and is completely independent, non-partisan and non-political. Its eleven members are appointed by the President from a list of nominees submitted by the National Academy of Sciences (NAS) which makes its nominations based solely on the expertise of the individual in relevant scientific and engineering disciplines. The independent technical peer review offered by the NWTRB contributes to the acceptance by the public and scientific communities for different approaches to managing nuclear

67

U.S. Fourth National Report-Joint Convention on the Safety of Spent Fuel Management and on the Safety of Radioactive Waste Management

waste. The NWTRB is an organization that performs an independent and integrated technical evaluation of DOE's implementation of the NWPA.

The NWTRB continues to conduct ongoing technical peer review of DOE efforts to implement the NWPA and for advising Congress and the Secretary of Energy. The NWTRB identifies technical issues and questions that should be addressed and creates unbiased technical information that can be used by Congress, the Secretary of Energy, or the Blue Ribbon Commission, to inform the evaluation of waste management alternatives and technologies. The NWTRB continues to report on the technical validity of DOE's efforts to Congress and the Secretary of Energy in twice-yearly reports, testimony, and correspondence.

E.2.4 State Regulatory Authorities

Provisions of law allow Federal agencies to delegate or relinquish certain regulatory responsibilities to the states having radioactive materials or nuclear facilities. More complex facilities such as nuclear power plants are regulated by Federal authorities. Regional arrangements allow closer coordination, such as using radioisotopes for medical uses. These arrangements are not necessarily mandatory, but, where the state can demonstrate adequate competencies, the appropriate Federal agency can transfer regulatory authority.

E.2.4.1 EPA Authorized States

EPA delegates authorities to states in two areas of radioactive waste management. NESHAPs regulations are based on the requirements of the Clean Air Act law, and the authority for delegating to states is described by law. A state must have emission limits at least as stringent as EPA's national standards. Most states have not pursued delegation of radionuclide NESHAPs. EPA's process for delegating RCRA hazardous waste requirements to states is similar. The state must have a program at least as stringent as EPA's, and the application for authorization must address specific areas of compatibility. The statutory basis for specific rules, however, may differ. Some "base" requirements must be adopted by states, while states may choose not to adopt other rules. The rule issued by EPA in 2001, and described in Section E.2.2.3, allowing mixed radioactive and hazardous waste generators to remain exempt from the hazardous waste requirements, for example, is not immediately effective in authorized states because it provides for a less stringent method of managing these wastes. For more information see Section E.2.7.1 of the Third National Report.

E.2.4.2 NRC Agreement States

The AEA, as amended, provides a statutory basis for NRC to relinquish to the states portions of its authority to license and regulate byproduct materials (radioisotopes); source materials (uranium and thorium); and certain quantities of special nuclear materials. As of December 2010, 37 of the 50 states have entered into Agreements with NRC. One additional state has filed intent to become an Agreement State.

Agreement States are those states having entered into an effective regulatory discontinuance agreement with NRC under subsection 274b. of the AEA. The role of the Agreement States is to regulate most types of radioactive material in accordance with the compatibility requirements of the AEA. These types of radioactive materials include source material (uranium and thorium), reactor fission byproducts, byproduct materials as defined in Section 11e. of the AEA, and quantities of special nuclear materials not sufficient to form a critical mass. NRC under its own internal practices periodically reviews the performance of each Agreement State to assure compatibility with its regulatory standards. See Section E.2.1.5 on NRC's IMPEP.

Agreement States issue radioactive material licenses, promulgate regulations, and enforce those regulations under the authority of each individual state's laws. The Agreement States conduct their licensing and enforcement actions under direction of the governors in a manner compatible with the licensing and enforcement programs of NRC.

NRC assistance to states entering into agreements includes review of requests from states to become Agreement States, or amendments to existing agreements, meetings with states to discuss and resolve NRC review comments, and recommendations for NRC approval of proposed agreements. NRC also conducts training courses and workshops; evaluates technical licensing and inspection issues from Agreement States; evaluates state rule changes; participates in activities conducted by the Organization of Agreement States[94] and the Conference of Radiation Control Program Directors, Inc.; and provides early and substantive involvement of the Agreement States in NRC rulemaking and other regulatory efforts. NRC also coordinates with Agreement States on event reporting and information, reciprocity arrangements, and responses to allegations reported to NRC involving Agreement States.

---

[94] See http://www.agreementstates.org for more information.

69

U.S. Fourth National Report-Joint Convention on the Safety of Spent Fuel Management and on the Safety of Radioactive Waste Management

F. GENERAL SAFETY PROVISIONS

Section F addresses general safety provisions in Articles 21-26 of the Joint Convention including:

- License holder responsibilities;
- Human and financial resources;
- Quality assurance;
- Operational radiation protection;
- Emergency preparedness; and
- Decommissioning.

This section also addresses Articles 4-9 and Articles 11-16. The following provisions are common for both spent fuel and radioactive waste management:

- General safety requirements;
- Existing facilities;
- Siting proposed facilities;
- Facility design and construction;
- Facility safety assessment; and
- Facility operation.

Sections G and H, address these same areas plus Articles 10 and 17 for Spent Fuel Disposal and Institutional Measures after Closure, and provide additional information specific to management of spent fuel or radioactive waste.

Section E presents the various regulations and directives, many of which are referenced in the following sections governing safety requirements in the U.S., including those for spent fuel management. Most of these regulations are available electronically on the internet. See Table A-2.

This report focuses on important issues and provisions mentioned above. For additional background and specific information, refer to Chapter F in the previous National Report.[95]

F.1 License Holder Responsibilities (Article 21)

The Joint Convention specifies each Contracting Party must ensure the prime responsibility for safety of spent fuel and radioactive waste management rests with the licensee, and each licensee takes the appropriate steps to meet its responsibility. The government has the responsibility only if there is no licensee. Nuclear Regulatory Commission (NRC) and Agreement State regulations ensure licensees are responsible for safe radioactive waste and spent fuel management. Commercial disposal facility licensees or operators will eventually transfer control of the site to Federal or state governmental agencies, which in turn will be responsible for protection of public safety and the environment.

---

[95] http://www.em.doe.gov/pdfs/3rdpercent20USpercent20Rptpercent20onpercent20SNFpercent20JC--percent20COMPLETEpercent20REPORTpercent20-percent2010percent2013percent2008.pdf.

F.2 Human and Financial Resources (Article 22)

Both commercial (NRC-regulated) and government (DOE) sectors have requirements to ensure human and financial resources are sustained for spent fuel and radioactive waste management activities. Table F-1 provides information from NRC on human resources in terms of full-time equivalent (FTE) staff dedicated to regulation in various programmatic areas. It should be noted that these are the resources requested in NRC's FY 2012 budget.

Table F-1 Distribution of NRC FY 2012 Budget Request Full-Time Equivalents in Staff	
Budgeted Program	**FTEs Requested for FY 2012**
Nuclear Reactor Safety	3032.9
Nuclear Materials and Waste Safety	868.5
Inspector General	58.0

Source: Nuclear Regulatory Commission

Table F-2 provides a breakdown for the total components for Nuclear Materials and Waste Safety including radioactive waste, spent fuel and oversight (which includes enforcement). The programmatic categories of nuclear materials and waste safety consist of 868.5 FTE. Approximately 295 FTEs or 34 percent of these FTEs are allocated to nuclear waste and spent fuel management.

Table F-2 NRC Staffing for Materials and Waste Management	
Regulatory Program[96]	**FTEs Requested for FY 2012**
Fuel Facilities	226.5
Nuclear Materials Users	347.1
High-Level Waste Repository	0
Spent Fuel Storage and Transportation	152.4
Decommissioning and Low-Level Waste	142.6
Materials and Waste Safety Total	868.5
Oversight (includes Enforcement)	**FTEs Requested for FY 2012**
Spent Fuel Storage and Transportation	23.6
Decommissioning and Low-Level Waste	27.3
Materials Users	99.5
Oversight Total	150.4

The NRC has a report, *NRC Summary of Performance and Financial Information, Fiscal Year 2010,*[97] which summarizes the performance of its mission to protect people and the environment through the regulation of nuclear power and use of nuclear material. It provides key financial and performance information for the U.S. Congress and the public to assess how well it has carried out its mission. This report also documents the finding of an independent auditor on NRC's condensed financial statements. This summary is

---

[96] Source: http://www.nrc.gov/reading-rm/doc-collections/nuregs/staff/sr1100/v27/.

[97] See http://www.nrc.gov/reading-rm/doc-collections/nuregs/staff/sr1542/v16/s1/sr1542v16s1.pdf.

72

U.S. Fourth National Report-Joint Convention on the Safety of Spent Fuel Management and on the Safety of Radioactive Waste Management

part of a larger effort to implement openness and transparency in NRC's program performance and financial management information.[98]

DOE field and program offices have primary regulatory functions at DOE through oversight of, and assuring compliance by, the management and operating contractors who manage and operate sites and facilities. Other DOE Headquarters organizations provide an additional layer of independent oversight. DOE's accounting and budgetary structure is different from NRC. Staffing for the specific DOE Headquarters organization is as follows: the Office of Health, Safety and Security (HSS) Office of Independent Assessment has a staff of 29 individuals assigned specifically to carry out environment, safety and health evaluations. The Office of Enforcement has a staff of nine at Headquarters who are engaged in safety, health and Price-Anderson enforcement. There are also Price-Anderson Coordinators in DOE Program Operations, Field and Area Offices. The DOE Chief of Nuclear Safety has a staff of eight who provide regulatory oversight. In the DOE National Nuclear Security Administration, there are four staff members in the Office of the Administrator and 10 under the Chief of Defense Nuclear Safety who provide regulatory oversight. In the DOE Environmental Management Office of Safety Management there is a staff of nine individuals who provide oversight of operations.

The Defense Nuclear Facility Safety Board had 103 FTEs in 2011.

F.2.1 Personnel Qualifications for NRC Licensees

NRC regulations require licensees to have qualified personnel. The requirements provide for an organizational structure of the licensee, both offsite and onsite, including lines of authority and assignments of responsibilities, whether in the form of administrative directives, contract provisions, or otherwise. NRC also has qualification requirements for its own personnel working on spent fuel and radioactive waste management regulatory activities.[99]

NRC establishes qualifications for licensees' employees responsible for operational safety and radiological health, including the radiation safety officer and health physics personnel. Technical qualifications include training and experience so the licensee's staff is competent to engage in the licensed activities. The licensee must additionally conduct a personnel training program and, within a plan, maintain an adequate complement of trained personnel to carry out licensed activities in a safe manner.

Operations of systems and components identified as important to safety must be performed only by trained and certified personnel or by personnel under the direct visual supervision of an individual with training and certification in such operation. Supervisory personnel directing operations important to safety must also be certified in such operations. Certain materials licensees must be qualified by training and experience to use the material for the purpose requested in such manner as to protect health and minimize danger to life and property.

---

[98] Additional detailed information on NRC's approach to open government is accessible at: http://www.nrc.gov/public-involve/open.html

[99] See Third U.S. National Report, Annex F-1 at: http://www.em.doe.gov/pdfs/3rd%20US%20Rpt%20on%20SNF%20JC--%20COMPLETE%20REPORT%20-%2010%2013%2008.pdf

73

U.S. Fourth National Report-Joint Convention on the Safety of Spent Fuel Management and on the Safety of Radioactive Waste Management

The physical condition and the general health of personnel certified for radioactive waste and spent fuel management operations important to safety may not be such as might cause operational errors endangering the public health and safety. Any condition potentially causing impaired judgment or motor coordination must be considered in the selection of personnel for activities important to safety. These conditions need not categorically disqualify a person, as long as appropriate provisions are made to accommodate the conditions.

F.2.2 DOE Qualification Requirements

DOE places requirements on contractors for training, proficiency testing, certification, and qualification of operating and supervisory personnel. Training requirements for nuclear safety management are in 10 Code of Federal Regulations (CFR) Part 830, and radiation worker protection in 10 CFR Part 835.

DOE directives further impose additional personnel training and qualification requirements for its activities. Developing and maintaining a technically competent workforce to accomplish its missions in a safe and efficient manner is accomplished through the Federal Technical Capability Program.[100] The objective of the Federal Technical Capability Program is to recruit, deploy, develop, and retain Federal personnel with the necessary technical capabilities to safely accomplish the Department's missions and responsibilities. The Department has identified guiding principles to accomplish that objective and identified four general functions of the Federal Technical Capability Program. The guiding principles and general functions are identified in DOE P 426.1[101] *Federal Technical Capabilities Policy for Defense Nuclear Facilities*, and DOE Order 426.2[102] *Personnel Selection, Training, Qualification, and Certification Requirements for DOE Nuclear Facilities*.

At a broader level, DOE is addressing future personnel needs through a comprehensive strategic plan to ensure its workforce can accomplish its mission in an efficient, effective, and productive manner for both business and safety operations. Please see DOE Strategic Human Capital Plan (FY 2006-2011)[103] for details.

F.2.3 Financial Surety

Licensees in the commercial sector must meet NRC requirements for financial surety. Spent fuel and radioactive waste management activities in the government sector (DOE facilities) have the financial assurance of the U.S. Government. Annual appropriations are made by the Congress. For nuclear materials facilities and activities regulated in NRC Agreement States, commercial licensees must meet the Agreement State financial assurance provisions.

The treatment of the trust funds for decommissioning a commercial power reactor differs depending on whether the licensee is a public utility or not. A public utility trust fund relies on funds collected in electric rates, and can take credit for future authorized collections. In addition, the public utility trust fund can take credit for projected earnings

---

[100] See: http://www.hss.doe.gov/deprep/ftcp/about.asp.
[101] See: https://www.directives.doe.gov/directives/current-directives/426.1-APolicy/view.
[102] See: https://www.directives.doe.gov/directives/current-directives/426.2-BOrder/view.
[103] See: http://humancapital.doe.gov/resources/DOEStrategicHumanCapitalPlan511.pdf.

74

U.S. Fourth National Report-Joint Convention on the Safety of Spent Fuel Management and on the Safety of Radioactive Waste Management

on the trust fund balance. The state public service commission determines the rate of return that can be used to calculate the earnings credit. Licensees who are not public utilities are called "merchant plants." They do not have prices guaranteed by a state public utility commission. Therefore, they cannot take credit for future collections. However, they are allowed to take credit for future earnings on the balance of the trust fund, at a rate no greater than 2 percent real rate of return. The period for projecting the earnings credit can extend up to approximately 60 years after permanent shutdown. However, if the licensee plans to take more than 7 years after permanent shutdown to complete decommissioning, it must provide additional financial assurance to cover the increased costs of a period of safe storage.

In addition to trust funds, a power reactor licensee may use several other methods to assure funding for decommissioning. These methods include guarantees, contractual obligations, and combinations thereof. If the power reactor licensee is a government entity, it may provide assurance by using a statement of intent to obtain funds for decommissioning from its governing legislative body. A power reactor licensee may propose other methods of assurance, but must show the method is equivalent to the methods listed in NRC's regulations in order to obtain approval. However, nearly all costs of power reactor decommissioning are assured using a trust fund.

F.2.3.1 Commercial LLW Management Facilities

The financial information provided by commercial LLW management facilities must be sufficient to demonstrate the financial qualifications of the applicant are adequate to carry out the activities for which the license is sought and other financial obligations. Each applicant must show it either possesses the necessary funds or has reasonable assurance of obtaining them to cover the estimated costs of conducting all licensed activities over the planned operating life of the project, including costs of construction and disposal.

Waste processors are subject to 10 CFR 20.1403 (or equivalent regulations in Agreement States). These regulations require sufficient financial assurance to enable an independent third party, including a governmental custodian, to implement responsibilities for control and maintenance of the site where the license is terminated with restrictions on future use. The financial assurance mechanism and amount are scrutinized by NRC before the license is terminated. No post-closure activities or institutional controls are needed for sites released free of future restrictions.

NRC financial assurance requirements for commercial LLW disposal facilities are contained in Subpart E of 10 CFR Part 61. Specific rules for funding of closure and stabilization are in 10 CFR 61.62 and for funding of institutional control in 10 CFR 61.63. A state has responsibility for review and acceptance of financial sureties in Agreement States in accordance with its regulations equivalent to Part 61.

The applicant must provide assurance that sufficient funds are available to carry out disposal site closure and stabilization (by an independent contractor, if necessary), including: (1) decontamination or dismantlement of structures; and (2) closure and stabilization of the disposal site so the need for active maintenance by the ultimate site owner is virtually eliminated and only minor custodial care, surveillance, and monitoring are required. The licensee's surety mechanism is reviewed annually by NRC (or

75

U.S. Fourth National Report-Joint Convention on the Safety of Spent Fuel Management and on the Safety of Radioactive Waste Management

Agreement State) to assure sufficient funds are available for completion of the closure plan.

To avoid unnecessary duplication – and if found adequate to accommodate NRC requirements – NRC accepts financial sureties consolidated with earmarked financial or surety arrangements established to meet requirements of other governing bodies for such decontamination, closure and stabilization.

The amount of surety changes with the predicted cost of future closure and stabilization. Factors affecting closure and stabilization cost estimates are summarized in 10 CFR 61.62(d). Financial surety arrangements generally acceptable to NRC include: surety bonds, cash deposits, certificates of deposit, deposits of government securities, escrow accounts, irrevocable letters or lines of credit, trust funds, and combinations of the above or other arrangements approved by NRC. Self-insurance, or any arrangement constituting pledging the assets of the licensee, does not satisfy the surety requirement for private sector applicants.

F.2.3.2   Spent Fuel and HLW Management Facilities

The Nuclear Waste Policy Act of 1982 (NWPA) requires utilities having a contract with the Department for the disposal of spent fuel or HLW to pay fees into the Nuclear Waste Fund sufficient to cover the costs associated with disposal activities for spent fuel and HLW. The fee, which is evaluated annually for sufficiency, currently is $0.001 per kilowatt-hour of nuclear power generated and sold.

Financial assurance for the storage of spent fuel is required under provisions at 10 CFR 72.22 for specifically licensed non-DOE independent spent fuel storage installations (ISFSIs) to ensure funds are available to store spent fuel in ISFSIs and for future decommissioning. ISFSI general licensees are covered by the financial assurance requirements in 10 CFR Part 50 for a reactor licensee. Financial mechanisms used include surety/insurance or other guarantee method, government statement of intent, or contractual obligations on the part of the firm's customers.

F.2.3.3   Uranium Recovery Waste Management Facilities

Financial surety arrangements must be established by each mill operator prior to the start of operations to assure sufficient funds will be available to carry out the Decontamination and Decommissioning (D&D) of the mill and site and for the reclamation of any tailings or waste disposal areas in the event the licensee is unable to do so. This process is similar for both conventional mills and in situ recovery operations (ISRs); the main difference is that ISRs have no tailings piles.

The amount of funds to be guaranteed by such surety arrangements must account for costs of an independent contractor for performing the work and must be based on NRC-approved cost estimates, which address:

- D&D of buildings and the site;
- Long-term site surveillance and control (if applicable); and
- Reclamation of tailings and/or waste areas in accordance with Appendix A to 10 CFR Part 40.

76

U.S. Fourth National Report-Joint Convention on the Safety of Spent Fuel Management and on the Safety of Radioactive Waste Management

Financial surety arrangements generally acceptable to NRC are surety bonds, trust funds, and letters of credit and combinations thereof or other arrangements approved by NRC. The surety must also cover payment of the charge for long-term surveillance and control at heap leach and conventional mill sites (long-term surveillance is not required at in situ recovery facilities). A minimum charge of $250000 (indexed to 1978 U.S. dollars) to cover the costs of long-term surveillance is paid by each mill operator to the General Treasury of the U.S. or to an appropriate state agency prior to the termination of a uranium or thorium mill license.

A variance in funding requirements for the long-term care charge may be specified by NRC if site surveillance or control requirements at a particular site are determined, on the basis of a site-specific evaluation, to be significantly greater than annual site inspections.[104] Eventual ownership of the uranium mill disposal site will be transferred to either DOE or an appropriate state agency for perpetuity. The funding should be adequate and should not impose costs to the long-term custodian during the long-term care period.

The surety is reviewed annually by NRC to recognize any increases or decreases resulting from inflation, changes in engineering plans, activities performed, and any other conditions affecting costs. This process yields a surety at least sufficient at all times to cover the costs of decommissioning and reclamation of the areas expected to be disturbed before the next license renewal.

F.2.3.4 Complex Material Sites Decommissioning

Many of the existing NRC-regulated decommissioning sites are complex and difficult to decommission for a variety of financial, technical, or programmatic reasons. These sites can be thought of as NRC "legacy" sites; those sites where past financial or operational events have created the problems needing a cleanup solution, and ultimately complete decommissioning and license termination. NRC evaluated the lessons from these existing legacy sites and has issued a final rule on decommissioning planning (See Section K.6) to change its current financial assurance and licensee operational requirements to minimize or prevent future legacy sites.[105] A detailed description of the decommissioning process for all materials sites is available at the NRC website.[106]

NRC experience applying the financial assurance regulations has resulted in many lessons learned for improving the regulations and reducing the risks to decommissioning financial assurance. NRC evaluated options for each of these funding risks and made recommendations for both existing and future material licensees. New recommendations from this process include using a risk-informed approach to identify high-risk operational indicators (e.g., spills, ground water contamination, and facility modification) and requiring updates to decommissioning cost estimates and financial assurance coverage.

The cost to decommission these facilities ranges broadly, from a few thousand up to the hundred million dollar range. More specific information on financial assurance for

---

[104] Conducted by the government agency responsible for long-term care of the disposal site to confirm its integrity and to determine the need, if any, for maintenance and/or monitoring, e.g., if fencing is necessary.
[105] Lessons learned from NRC's experiences in decommissioning can be accessed at http://www.nrc.gov/about-nrc/regulatory/decommissioning/lessons-learned.html.
[106] See http://www.nrc.gov/about-nrc/regulatory/decommissioning/process.html

77

U.S. Fourth National Report-Joint Convention on the Safety of Spent Fuel Management and on the Safety of Radioactive Waste Management

decommissioning, such as the need for licensees to provide a decommissioning funding plan, is provided at http://www.nrc.gov/about-nrc/regulatory/decommissioning/finan-assur.html.

NRC regulations at 10 CFR 20.1406, *Minimization of Contamination*, specifically require that new applications describe how design and operations will minimize contamination and facilitate eventual decommissioning. As mentioned above, NRC has published a final rule on decommissioning planning. One aspect of the rule focuses on ensuring that licensees have adequate financial assurance to complete decommissioning, while the other ensures that licensees have an adequate ground water monitoring program in place and will implement measures to minimize ground water contamination. NRC affirmed the final rule on November 30, 2010, and it takes effect in January 2012; except for some provisions to take effect in March 2013. Waste minimization is more fully discussed in Section F.7.2.

Additional details on financial assurance in decommissioning material sites can be found in Consolidated Decommissioning Guidance, Financial Assurance, Recordkeeping, and Timeliness (NUREG-1757, Vol. 3).[107]

F.3 Quality Assurance (Article 23)

The following subsections provide a summary of quality assurance (QA) requirements prescribed by NRC and DOE for spent fuel and waste management activities. QA requirements apply to licensees, licensed subcontractors, DOE contractors and subcontractors, and to suppliers. QA programs are applied to design, purchase, fabrication, handling, shipping, storing, cleaning, assembly, inspection, testing, operation, maintenance, and repair, and modification of structures, systems and components important to safety.

For commercial licensees, NRC generally inspects the facilities and activities irrespective of organizational affiliation. NRC holds the organization specified in the license as the responsible party for enforcement purposes. Similarly, DOE holds its prime contractors responsible for all subcontractors in addition to direct enforcement authority over subcontractors under the Price-Anderson Act.[108]

F.3.1 NRC Quality Assurance

Specific technical information needed for adequate demonstration of the performance objectives and the applicable technical requirements for a LLW disposal operation, include a description of the quality assurance program, tailored to LLW disposal, developed and applied by the applicant for the determination of natural disposal site characteristics and for quality assurance during the design, construction, operation, and closure of the land disposal facility and the receipt, handling, and emplacement of waste. [10 CFR 61.12(j)] Guidance to applicants on how to meet the QA regulatory requirements in Part 61 is also provided in NUREG-1293, Revision 1, *Quality Assurance Guidance for a Low-Level Radioactive Waste Disposal Facility (April 1991)*.[109] The

---

[107] See http://www.nrc.gov/reading-rm/doc-collections/nuregs/staff/sr1757/v3/index.html.

[108] For further information on this subject, refer to the U.S. Atomic Energy Act, Sections 234a and 170d.

[109] Available from the NRC Web-based ADAMS at http://www.nrc.gov/reading-rm/adams.html.- ADAMS Accession Number ML11242A180

78

U.S. Fourth National Report-Joint Convention on the Safety of Spent Fuel Management and on the Safety of Radioactive Waste Management

criteria in NUREG-1293 are similar to the criteria contained in 10 CFR Part 50, Appendix B, *Quality Assurance Criteria for Nuclear Power Plants and Fuel Reprocessing Plants*. Although Appendix B to 10 CFR Part 50 is not applicable to the NRC's LLW disposal regulation, the criteria it contains are basic to any nuclear regulatory QA program. For example, quality assurance includes quality control and is implemented by inspections, audits, calibration testing, as well as review of design and operational protocols. Other sources of guidance include: NUREG-1383, *Guidance on the Application of Quality Assurance for Characterizing a Low-Level Radioactive Waste Disposal Site: Final Report* (1990),[110] which provides QA guidance related to site characterization activities. Chapter 9 of both *Standard Format and Content of a License Application for a Low-Level Radioactive Waste Disposal Facility* (NUREG-1199) and *Standard Review Plan for the Review of a License Application for a Low-Level Radioactive Waste Disposal Facility* (NUREG-1200) provide additional QA guidance for potential 10 CFR Part 61 applicants, *QA Guidance for Low-level Waste Disposal Facilities*.

The QA program for storage of spent fuel, HLW, and reactor-generated Greater-than-Class C Low-Level Waste (GTCC LLW) is described in 10 CFR Part 72, Subpart G. An additional useful document is NUREG/CR-6314, *Quality Assurance Inspections for Shipping and Storage Containers.*[111] The QA requirements for packaging and transportation of licensed radioactive material are provided in Subpart H of 10 CFR Part 71. This would include packaging and transport of radioactive sources – disused or otherwise – unless handled under an exemption provision. NRC Regulatory Guide 7.10 describes QA programs for packaging for transport of radioactive material.[112]

Quality assurance is generally addressed as part of the license requirements for uranium recovery operations. Areas where quality assurance is particularly important include: disposal cell performance; monitoring, injection, and recovery well construction; and final cover system construction. Typically, technical specifications are developed to provide requirements for materials and construction techniques. A quality assurance, testing program for the operational phases including supervision by a qualified engineer or scientist, is established to assure the specifications are met. In some cases, onsite pilot projects on a smaller scale can provide a demonstration that reclamation and restoration strategies are achievable.

Quality Assurance plays a significant role in decommissioning of nuclear facilities. Decommissioning plans include a quality assurance program to determine if the licensee has adequate controls in place to support the decommissioning. The QA program should address document control, control of measuring and test equipment, corrective action, QA records, audits, environmental monitoring, instrumentation and surveillances. Requirements are specified in 10 CFR 72 Subpart G, 10 CFR 50.82, and guidance is available from NUREG-1757 *Consolidated Decommissioning Guidance: Decommissioning Process for Materials Licensees*.

---

[110] http://www.osti.gov/bridge/servlets/purl/6503151-A8yRkR/6503151.pdf

[111] See http://www.nrc.gov/reading-rm/doc-collections/nuregs/contract/cr6314/cr6314.pdf.

[112] Quality Assurance guidance for Part 71 is provided in the Regulatory Guide 7.10, *Establishing Quality Assurance Programs for Packaging Used in Transport of Radioactive Material,* Revision Number 2.

79

U.S. Fourth National Report-Joint Convention on the Safety of Spent Fuel Management and on the Safety of Radioactive Waste Management

F.3.2 DOE Quality Assurance

Most DOE activities are subject to QA requirements found in DOE orders and guidance. DOE QA requirements are specified at 10 CFR 830.120. DOE programs must implement the QA criteria to achieve adequate protection of the workers, the public, and the environment, taking into account the work to be performed and its hazards. They must develop their QA programs by applying ten QA criteria using a graded approach. The ten QA criteria fall within three areas: management, performance, and assessment. The management criteria are QA program, personnel training and qualification, quality improvement, documents and records. The performance criteria are work processes, design, procurement and inspection, and acceptance testing. The assessment criteria are management assessment and independent assessment. The QA program plan must describe how the criteria are satisfied and how the graded approach is applied. DOE performs internal audits and assesses whether its contractors have satisfactorily implemented the DOE quality assurance program.

F.4   Operational Radiation Protection (Article 24)

The following sections describe radiation protection responsibilities at EPA, NRC, and DOE. The U.S. Government also has access to leading experts in radiation protection through institutions such as the National Academy of Sciences (NAS)/National Research Council and the NCRP. The NAS is a private, nonprofit institution providing science, technology and health policy advice under a Congressional charter. The NAS established a Board on Radioactive Waste Management (now part of the new Nuclear and Radiation Studies Board) focusing on waste management and disposal.

F.4.1   U.S. Environmental Protection Agency

EPA is responsible for issuing guidance to Federal agencies on radiation protection matters. EPA provides emergency response training and analytical support to state, local, and tribal governments and works closely with other national and international radiation protection organizations to further our scientific understanding of radiation risks.

Primary radiation protection regulations for spent fuel management include 40 CFR Part 190, *Environmental Radiation Protection Standards for Nuclear Power Operations*, and 40 CFR Part 191, *Environmental Radiation Protection Standards for Management and Disposal of Spent Nuclear Fuel, High-level and Transuranic Radioactive Wastes*.

Another radiation protection regulation related to 40 CFR Part 191, pertaining to transuranic radioactive waste (not spent fuel) management at DOE's Waste Isolation Pilot Plant (WIPP) geologic repository, is found in 40 CFR Part 194, *Criteria for the Certification and Re-certification of the Waste Isolation Pilot Plant's Compliance with 40 CFR Part 191 Disposal Regulations*. See Section E.2.2.1 for additional information.

Federal guidance is a set of guidelines developed by EPA, for use by Federal agencies responsible for protecting the public from the harmful effects of radiation. EPA's Federal guidance is often applied by state agencies and the private sector to ensure consistency of practice. Guidance documents produced by EPA are available on the internet.[113]

---

[113] EPA Radiation Protection Program, http://www.epa.gov/radiation/programs.html.

Specific dose limits are provided in Table F-3. Some key radiation protection guidance documents are listed in Annex F-1.

F.4.2 NRC General Radiological Protection Limits

The provisions for general safety for workers and protection of the public during the operational phase of commercial radioactive waste management facilities are addressed in NRC regulations contained in 10 CFR Part 20, *Standards for Protection Against Radiation*. 10 CFR Part 20 includes agency requirements for:

- Dose limits for radiation workers and members of the public;
- Monitoring and labeling radioactive materials;
- Posting radiation areas; and
- Reporting the theft or loss of radioactive material.

The provisions in 10 CFR Part 20 also include:

- Penalties for not complying with NRC regulations; and
- Tables of individual radionuclide exposure limits.

NRC regulates commercial nuclear power generation as well as medical, academic, and industrial uses of radioactive material.[114] NRC promulgates safety regulations expressed in annual total effective dose equivalents (TEDEs),[115] as well as air and liquid effluent release concentrations. See Table F-3 for occupational and public health protection standards.

F.4.2.1 Occupational Dose Limits

Operations are conducted so the occupational dose to individual adults complies with the appropriate annual limit; refer to Table F-3. Annual occupational dose limits are established at 10 CFR 20.1201 for adults and Section 20.1207 for minors. There are other specific conditions, such as for planned special exposures and specific organ limits, as well as considerations for a soluble uranium chemical toxicity intake limit of 10 milligrams in a week. For more specific constraints refer to 10 CFR Part 20, Subpart B.

NRC maintains the Radiation Exposure Information and Reporting System for Radiation Workers, which provides the latest available information on radiation exposure to the workforce at certain NRC-licensed facilities. It also contains information concerning the recording and reporting requirements of NRC licensees. This information and other details on occupational exposure are available on the internet and updated annually.[116]

---

[114] NRC can relinquish regulatory authority for nuclear materials to Agreement States under the provisions of the Atomic Energy Act, as amended. Refer Section E.2.4.2. for more specific information.

[115] Dose is defined here as the total effective dose equivalent, which is defined as the sum of the deep-dose equivalent for external exposures and the committed effective dose equivalent for internal exposures.

[116] NRC's Radiation Exposure Information and Reporting System (REIRS) for Radiation Workers is accessible at http://www.reirs.com.

81

U.S. Fourth National Report-Joint Convention on the Safety of Spent Fuel Management and on the Safety of Radioactive Waste Management

An example of this information is provided in Figure F-1 that provides historical TEDEs for ISFSIs.[117]

F.4.2.2 Public Dose Limits

Public individual dose limits are provided in Table F-3. These dose limits are exclusive of the contributions from background radiation, any medical administration to individuals, and other contributions not attributable to other regulated operations.

Exposure limits for specific situations are provided in 10 CFR Part 20.

F.4.2.3 Radiological Criteria for License Termination (Decommissioning)

For protection of the public, dose-based requirements for licensees seeking license termination are found in 10 CFR Part 20, Subpart E (see Table F-3). These regulations establish two final states for licensee termination: unrestricted use and restricted use. In addition to the specific limits for each state, NRC requires licensees to maintain doses as low as reasonably achievable (ALARA). This means the licensee must make every reasonable effort to reduce the dose as far below the specified limits as is practical, taking into account the state of technology and economics. See 10 CFR 20.1003.

F.4.2.4 LLW Disposal Sites

Protection of the general population from releases of radioactivity from a LLW disposal facility is also dose-based. Reasonable efforts should be made to maintain releases of radioactivity in effluents to the general environment ALARA. See Table F-3 for specific applications of dose.

F.4.2.5 Uranium Mill Tailings Disposal Sites

Reclaimed uranium mills are required to meet a radon release constraint in 10 CFR Part 40, Appendix A in addition to the annual public and occupational dose limits described in the previous section. See Table F-3. The 0.7 Bq/m2-s radon release from uranium mill tailings was based on the cost-effectiveness of control for a thick earthen cover design, taking into consideration individual and population doses. There are also groundwater concentration limits for radionuclides and certain hazardous constituents, as well as a design control requirement.

---

[117] This document is published as NUREG-0713, *Occupational Radiation Exposure at Commercial Nuclear Power Reactors and Other Facilities* 2009, Vol. 31, NRC, June 2010, available at http://www.reirs.com/nureg2009/nureg2009.pdf.

Average Measurable Dose/per Worker (10⁻² Sieverts)

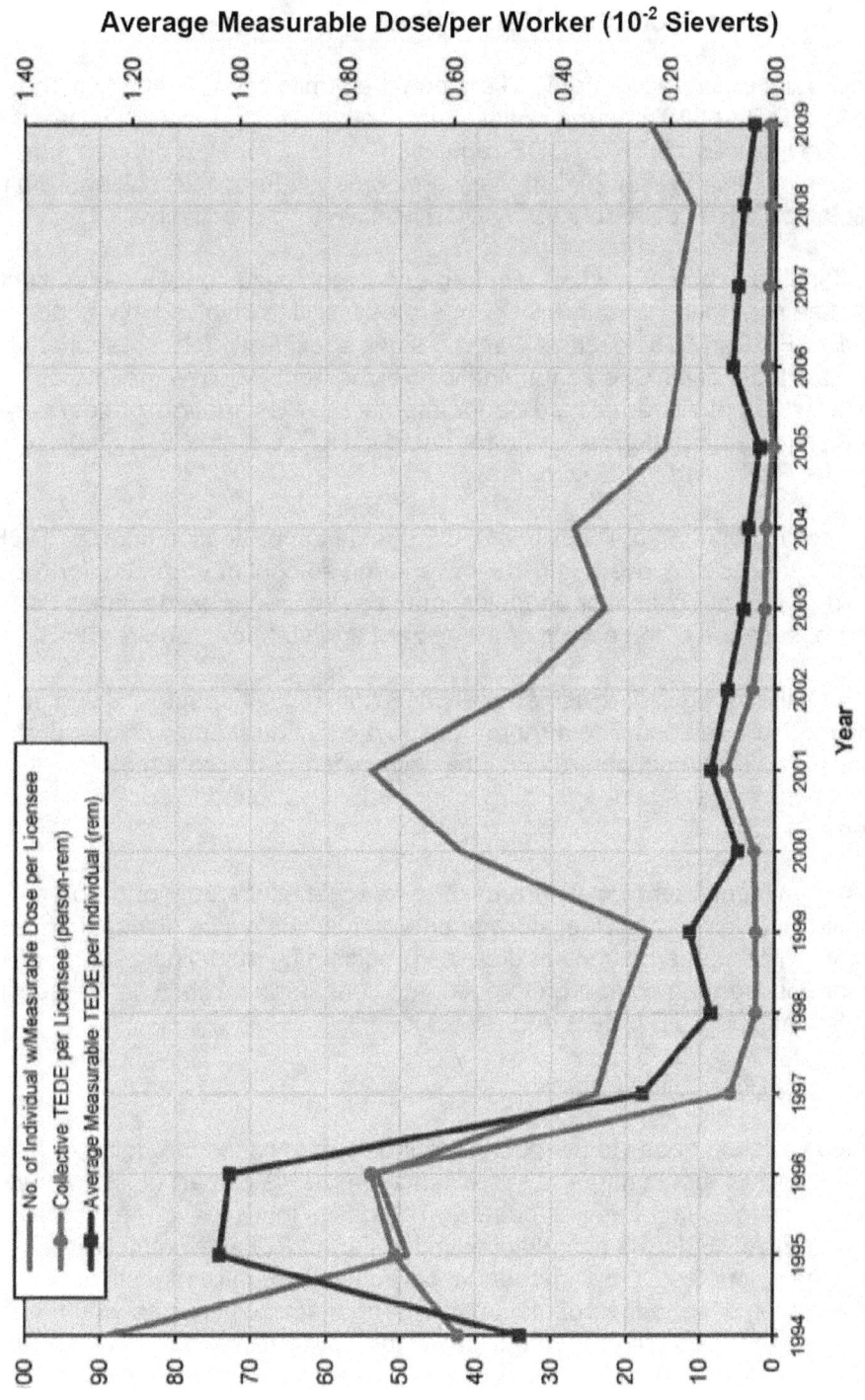

Average Number of Workers with Measurable Dose, Collective TEDE per Licensee (10⁻² Sieverts)

Figure F-1 Average Annual Values at Independent Spent Fuel Storage Facilities, 1994-2009

F.4.3 DOE Radiation Protection Regulations

DOE requires radiation protection for workers and the public in its regulations and directives. 10 CFR Part 835 governs radiation protection of workers at DOE facilities and activities not licensed by NRC. DOE regulations in 10 CFR Part 835 are similar to NRC regulations in 10 CFR Part 20, but there are some differences resulting from the types of radiological activities regulated by DOE and NRC, respectively.

DOE occupational radiation protection requirements emphasize contamination control and internal dose monitoring because DOE operates facilities involved in weapons production. 10 CFR Part 835 specifies warning signs specifically for contamination areas, contains a table of surface contamination values, and requires the use of bioassay data instead of air sampling data for internal dose estimation in most cases. Further radiation protection requirements are found in DOE Order 458.1, *Radiation Protection of the Public and the Environment*.

There is no time limit on the applicability of DOE's radiation protection criteria. DOE considers whether risks may eventually be low enough so continued protection would not be needed. DOE, or successor agencies, may be required in some cases, to maintain control because of the nature of the hazard and statutory requirements.

Compliance with these regulations is generally determined by inspectors using survey equipment to measure radionuclide airborne or liquid concentrations within and at control boundaries. These concentrations are determined to be representative of TEDEs or of effective doses corresponding to individuals exposed to such concentrations.

Safety assessment computer models are used to forecast exposures, prior to operating a nuclear facility, including spent fuel storage and radioactive waste disposal on a predictive basis. The concentrations and doses predicted by modeling a range of potential scenarios are then compared to dose and concentration limits in the applicable Federal regulations and DOE orders and manuals.

F.4.3.1 Collective Dose to the Public

DOE estimates radiation doses to the public around its many sites through extensive continuous radiological monitoring and surveillance programs as part of its commitment to communities where its facilities are located. The offsite individual doses remain well below DOE and EPA NESHAPs compliance limits. Figure F-2 shows the historical trend. To put the estimated DOE-wide annual collective dose in perspective, background radiation dose to the population in a large metropolitan area would be more than 20000 person-Sv (two million person-rem) annually, from natural and man-made sources

84

U.S. Fourth National Report-Joint Convention on the Safety of Spent Fuel Management and on the Safety of Radioactive Waste Management

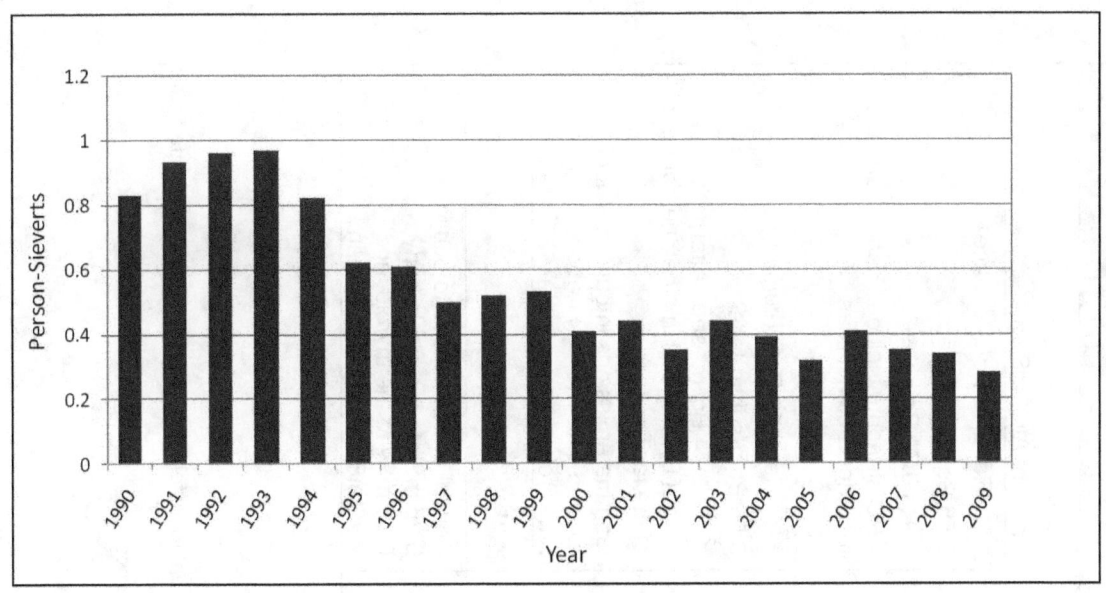

Figure F-2 Estimated Off-Site Radiation Dose to the Public around DOE Sites

F.4.3.2 Dose to DOE Workers

DOE keeps radiation exposures to workers ALARA within the constraints imposed by work, equipment, and technical conditions. Only 14 percent (11287 out of 83208) of DOE workers monitored for radiation dose received a measurable dose in 2008. In 2008, the average annual measurable dose to a worker was 0.61 mSv (61 mrem), and the collective dose was 6.90 person-Sv (690 person-rem). One individual exceeded the 50 mSv (5 rem) annual limit in 2007 (receiving 75 mSv (7.5 rem)). In 2007 and 2008 one individual in each year exceeded the 20 mSv (2 rem) administrative limit.

To place DOE worker dose in perspective, the average American receives approximately 6.2 mSv/a (620 mrem/yr) from natural and man-made sources.[118] The majority of DOE workers with a measurable dose in 2008 (9341 out of 11287) received less than 1 mSv (100 mrem) total effective dose equivalent. Thousands of people work in radiation areas every day without receiving significant radiation exposure, which may be attributable to the effectiveness of ALARA controls.

---

[118] Source: *Ionizing Radiation Exposure of the Population of the U.S.*, National Council on Radiation Protection and Measurements Report No. 160, 2009.

Table F-3 Major Radiation Protection Standards

Summary Table Regulation	Agency	US Standard Limit	SI Equivalent
General Public (10 CFR 20.1301 & DOE Order 458.1)	DOE & NRC	Total Effective Dose Equivalent (TEDE[119]): 100 mrem/year	TEDE 1 mSv/a Ambient Air 1 mSv/a[120]
Uranium mill tailings (40 CFR Part 61 and 192 & 10 CFR Part 40 App. A)	DOE, EPA & NRC	$^{226/228}$Ra: 5 pCi/g (surface) 15 pCi/g (subsurface) ^{222}Rn 20 pCi/m^2-sec. NRC standard includes benchmark dose for other radionuclides Public dose per 40 CFR Part 190	$^{226/228}$Ra: 0.19 Bq/g (surface) 0.56 Bq/g(subsurface) ^{222}Rn 0.74 Bqi/m^2-sec NRC standard includes benchmark dose for other radionuclides
Residual radioactive material UMTRCA Title I Facilities (40 CFR Part 192)	DOE & EPA	$^{226/228}$Ra: 5 pCi/g (surface) 15 pCi/g (subsurface) ^{222}Rn 20 pCi/m^2-sec. Alternatively, a limit of 0.5 pCi/l in air at the boundary of the facility Gamma emissions in buildings not to exceed 20 microroentgen/hr.above background	$^{226/228}$Ra: 0.19 Bq/g (surface) 0.56 Bq/g(subsurface) ^{222}Rn 0.74 Bqi/m^2-sec. Alternatively, a limit of 0.0185 Bq/l in air at the boundary of the facility. Gamma emissions in buildings not to exceed 5.16 mC/kg-hr above background.
High-level waste operations (10 CFR Part 60)	DOE & NRC	100 mrem/year	1 mSv/a
Commercial Low-level waste disposal (10 CFR Part 61)	NRC	Annual effective dose to public 25 mrem to the whole body 75 mrem to the thyroid, and 25 mrem to any other organ	Annual effective dose to public 0.25 mSv to the whole body 0.75 mSv to the thyroid, and 0.25 mSv to any other organ

[119] Dose is defined here as the total effective dose equivalent, which is defined as the sum of the deep-dose equivalent for external exposures and the committed effective dose equivalent for internal exposures.

[120] Limit is exclusive of radiation from background, medical administrations, and contributions from non-licensed sources

Table F-3 Major Radiation Protection Standards

Summary Table Regulation	Agency	US Standard Limit	SI Equivalent
DOE Low-level waste disposal (DOE Order 435.1)	DOE	TEDE to public: 25 mrem/year from all exposure pathways, excluding the dose from radon and its progeny in air.	TEDE to public: 0.25 mSv/year from all exposure pathways, excluding the dose from radon and its progeny in air.
		TEDE to public via the air pathway does not exceed 10 mrem/year, excluding the dose from radon and its progeny.	TEDE to public via the air pathway does not exceed 0.10 mSv/year, excluding the dose from radon and its progeny.
		Release of radon is less than an average flux 20 pCi/m^2/s at the surface of the disposal facility; alternatively, a limit of 0.5 pCi/l in air at the boundary of the facility.	Release of radon is less than an average flux of 0.74 Bq/m^2/s at the surface; alternatively, a limit of 0.0185 Bq/l in air at the boundary of the facility.
Effluent emissions (10 CFR Part 20)	NRC	Radionuclide specific activities, in Appendix B => 50 mrem/year	Radionuclide specific activities, in Appendix B => 0.5 mSv/a
Drinking water (40 CFR Part 141)	DOE, EPA & NRC	Radium: 5 pCi/L Gross Alpha 15 pCi/L (excludes Rn & U) Beta/photon: 4 mrem/year Uranium: 30 µg/L	Radium: 0.19 Bq/L Gross Alpha 0.56 Bq/L (excludes Rn & U) Beta/photon: 0.04 mSv/a Uranium: 30 µg/L
Uranium fuel cycle (40 CFR Part 190)	DOE, EPA & NRC	25/75/25 mrem/year	0.25/0.75/0.25 mSv/a
Air emissions (National Emission Standards for Hazardous Air Pollutants) (40 CFR Part 61, H)	DOE, EPA & NRC	10 mrem/year to nearest off-site receptor	0.1 mSv/a to nearest off-site receptor

Table F-3 Major Radiation Protection Standards

Summary Table Regulation	Agency	US Standard Limit	SI Equivalent
Superfund (CERCLA) cleanup (40 Part CFR 300)	DOE, EPA & NRC	Protective of human health & environment (lifetime risk), Complies with Applicable or Relevant and Appropriate Requirements (ARARs)	Not Applicable
Decommissioning (10 CFR Part 20, Subpart E)	NRC	Unrestricted Use: 25 mrem/yr TEDE plus ALARA Restricted Use: If institutional controls fail, not to exceed 100 mrem/yr or 500 mrem/yr.	Unrestricted Use: 0.25 mSv/a TEDE plus ALARA Restricted Use: If institutional controls fail, not to exceed 1 mSv/a or 5 mSv/a.
Occupational standards (DOE 10 CFR Part 835 OSHA 29 CFR 1910.1096 NRC 10 CFR Part 20.1201)	DOE, OSHA & NRC	5000 mrem/year & ALARA	50 mSv/a & ALARA
Spent Fuel, HLW, TRU management and disposal (40 CFR Part 191)	DOE, EPA & NRC	Annual Dose to Any Member of the Public from Management and Storage: *NRC-licensed sites* 25 mrem whole body 75 mrem thyroid 25 mrem other organ *DOE disposal sites (non-NRC licensed)* 25 mrem whole body 75 mrem thyroid Disposal Standards Applicable for 10,000 Years After Disposal: 15 mrem/year committed effective dose to any member of the public; Radionuclide-specific release limits to the accessible environment; Ground-water concentrations not to exceed drinking water limits	Annual Dose to Any Member of the Public from Management and Storage: *NRC-licensed sites* 0.25 mSv whole body 0.75 mSv thyroid 0.25 mSv other organ *DOE disposal sites (non-NRC licensed)* 0.25 mSv whole body 0.75 mSv thyroid Disposal Standards Applicable for 10,000 Years After Disposal: 0.15 mSv/year committed effective dose to any member of the public; Radionuclide-specific release limits to the accessible environment; Ground-water concentrations not to exceed drinking water limits

Table F-3 Major Radiation Protection Standards

Summary Table Regulation	Agency	US Standard Limit	SI Equivalent
Spent Fuel and HLW at Yucca Mountain (40 CFR Part 197 & 10 CFR Part 63)	DOE, EPA & NRC	15 mrem/year to any member of the public from storage;	0.15 mSv/year to any member of the public from storage;
		15 mrem/year to the Reasonably Maximally Exposed Individual (RMEI) for 10000 years after disposal for undisturbed performance and from human intrusion;	0.15 mSv/year to the Reasonably Maximally Exposed Individual (RMEI) for 10000 years after disposal for undisturbed performance and from human intrusion;
		100 mrem/year to the RMEI between 10000 and 1 million years after disposal for undisturbed performance and from human intrusion;	1 mSv/year to the RMEI between 10000 and 1 million years after disposal for undisturbed performance and from human intrusion;
		Ground-water concentrations not to exceed drinking water limits for 10000 years after disposal	Ground-water concentrations not to exceed drinking water limits for 10000 years after disposal
Independent Spent Fuel Storage Installations (ISFSIs) (10 CFR Part 72)	NRC	Annual dose equivalent to a real individual member of the public	Annual dose equivalent to a real individual member of the public
		25 mrem to the whole body	0.25 mSv to the whole body
		75 mrem to the thyroid, and	0.75 mSv to the thyroid, and
		25 mrem to any other organ	0.25 mSv to any other organ

Note: U.S. Standards for off-site transportation are excluded from the scope of the Joint Convention, and therefore not listed.

F.4.4 Other Radiation Protection Regulations

EPA has the prime role in setting U.S. radiation protection standards that are implemented by NRC, DOE, and other Federal agencies:

- The Occupational Health & Safety Administration of the Department of Labor (DOL) has regulations dealing with worker protection from ionizing radiation found in 29 CFR; and
- The Mine Safety and Health Administration of the DOL has safety and health regulations related to underground mining in 30 CFR Part 57, subparts 4037 to 5047.

Limits for air and water discharges from spent fuel/radioactive waste facilities are established through rulemaking by the responsible agency. EPA has issued rules for spent fuel, HLW, transuranic (TRU) waste, commercial nuclear fuel cycle, and uranium/thorium mill tailings facilities. See EPA's responsibilities throughout Section E.

NRC implements these rules and has established rules for commercially generated low-level radioactive waste facilities. NRC collects data from all nuclear power plant (NPP) licensees and reports are submitted twice each year in accordance with Regulatory Guide 1.21 Measuring, Evaluating, and Reporting Radioactive Material in Liquid and Gaseous Effluents and Solid Waste. Similarly, fuel cycle facility licensees are required to submit semi-annual reports in accordance with Regulatory Guide 4.16 Monitoring and Reporting Radioactivity in Releases of Radioactive Materials in Liquid and Gaseous Effluents from Nuclear Fuel Processing and Fabrication Plants and Uranium Hexafluoride Production Plants. Commercial NPP licensees produce annual reports of their discharges; information on environmental monitoring is summarized on the NRC website.[121]

Information on radiological discharges from DOE facilities engaged in waste management, environmental clean-up and spent fuel activities is available through DOE Annual Site Environmental Reports (ASERs).[122]

DOE regulates air and water discharges from its radioactive waste facilities through its internal orders, while airborne emissions from DOE facilities are regulated by EPA.

Many states have comprehensive radiation control programs. These programs, for example, may regulate the use of diagnostic and therapeutic x-ray equipment and certain radioactive materials or conduct environmental monitoring.

F.5 Emergency Preparedness (Article 25)

Article 25 specifies spent fuel and radioactive waste management facilities must have appropriate on-site and, if necessary, off-site emergency plans, and should be tested at an appropriate frequency. Additionally, Article 25 requires each Contracting Party to

---

[121] Information on operating reactor experience, including releases to the environment, are located at http://www.nrc.gov/reactors/operating/ops-experience.html. In addition, 10 CFR 72.44(d)(3) requires the annual submittal of a summary of effluents from the independent spent fuel storage facility. All reports can also be found in ADAMS at the NRC public website: http://www.NRC.gov.

[122] The ASERs are available at: http://www.hss.energy.gov/nuclearsafety/env/reports/aser/aserlinks.pdf.

prepare and test emergency plans on its territory in the event of a radiological emergency at a spent fuel or radioactive waste management facility in the vicinity of its territory. The following subsections describe the extensive emergency preparedness and emergency management programs in place at NRC-licensed and DOE facilities.

F.5.1 NRC Emergency Preparedness

F.5.1.1 Nuclear Facility Response Plans

NRC regulations require comprehensive emergency plans be prepared and periodically exercised to assure actions are taken to notify and protect citizens in the vicinity of a nuclear facility during an emergency. Although nuclear power plants, as well as fuel fabrication and uranium conversion and enrichment facilities, have active components potentially requiring immediate protective response to mitigate the effects of an accident or a terrorist attack, radioactive waste disposal systems are passive. For radioactive waste management and spent fuel management at a nuclear power plant or other significant nuclear fuel cycle facility, the emergency preparedness program is modified by license condition upon the facility's entry into the decommissioning phase. The revised provisions for emergency preparedness and response will be modified commensurate with the hazard of the materials remaining within the former controlled areas.

The vast majority of events reported to NRC are routine and do not require activation of its incident response program. See *About Emergency Response* [123] for information on how NRC responds to threatening public health and safety emergencies. NRC Regulatory Guide 3.67 [124] provides information on the classification of emergencies as either "alerts" or "site area emergencies" for nuclear material and fuel cycle facility licensees. Some nuclear materials licensees may also use the "Unusual Event" classification to notify officials of events of lower safety significance, although not required by NRC regulations. [125] DOE also has published classification guidance in DOE G151.1-1, *Categorization and Classification of Operational Emergencies.* [126]

Although the severity and extent of hazards associated with spent fuel or radioactive waste management facilities are different than those associated with a nuclear power plant, many of the elements for emergency response are still applicable. An Emergency Plan is required by NRC regulations in any license application for a spent fuel storage facility not located on the site of an operating nuclear reactor. There must be semiannual communications checks with offsite response organizations and biennial onsite exercises to test response to simulated emergencies. Other radiological/health physics, medical, and fire drills are conducted annually. There are similar requirements to have an Emergency Plan for materials licenses using large quantities of radioactive materials in unsealed form, on foils or plated sources, or sealed in glass.

NRC has no plans to change the emergency classification system for commercial nuclear power plants. NRC does not directly require a "General Emergency" category for non-reactor Fuel Cycle Facilities and other materials facilities. The Emergency

---

[123] See http://www.nrc.gov/about-nrc/emerg-preparedness/respond-to-emergency.html.
[124] See http://pbadupws.nrc.gov/docs/ML1033/ML103360487.pdf.
[125] Further information on the NRC's emergency event classification is available at:
http://www.nrc.gov/about-nrc/emerg-preparedness/about-emerg-preparedness/emerg-classification.html.
[126] See http://www.directives.doe.gov/.

Planning and Community Right-to-Know Act requires off-site "general emergency" planning for facilities possessing hazardous chemicals (including toxic uranium) on-site. NRC simply requires certification that these requirements have been met. It should also be noted "General Emergency" planning (for off-site areas) is much broader in scope, detail, complexity and politics than "Site Area" emergency planning.

F.5.1.2 Emergency Response

NRC activates its incident response program at its Headquarters Operations Center and one of its four Regional Incident Response Centers (Region I in King of Prussia, Pennsylvania; Region II in Atlanta, Georgia; Region III, in Lisle, Illinois; and Region IV in Arlington, Texas) in response to an event at an NRC-licensed facility potentially threatening public health and safety, or the environment. NRC's highest priority is to provide expert consultation, support, and assistance to state and local public safety officials responding to the event. Teams of specialists are assembled at the Headquarters Operations Center and Regional Incident Response Center to obtain and evaluate event information and to assess the potential impact of the event on public health and safety and the environment once NRC's incident response program is activated. Scientists and engineers analyze the event and evaluate possible recovery strategies. Other experts evaluate the effectiveness of protective actions recommended by the licensee and implemented by state and local officials to minimize the impact on public health and safety and the environment. Communications with the news media, state, other Federal agencies, the Congress, and the White House, and potentially affected international counterparts are coordinated through the Headquarters Operations Center.

Communications with the media remain the responsibility of NRC Office of Public Affairs. Personnel, both in the Operations Center and in the Headquarters Office of Public Affairs, will communicate frequently with the media via press releases, Web postings (and a special emergency event Web page), press conferences, if needed, and an open media-bridge telephone line. NRC regional public affairs officers normally interact with media personnel during an event that does not result in a change to NRC's response mode or if the agency is in the Monitoring Mode. Additional information on effective risk communication by NRC is provided at: http://www.nrc.gov/reading-rm/doc-collections/nuregs/brochures/br0308/.

NRC's role, as well as the roles of other Federal agencies in the coordinated emergency response to a nuclear accident, is described in the Nuclear/Radiological Incident Annex of the National Response Framework.[127] NRC will immediately dispatch a team of experts from the Regional Office to the site if event conditions warrant.

F.5.1.3 Emergency Response Exercises

NRC Headquarters and Regional staff members typically participate in several emergency response exercises each year for materials facilities. This is in addition to participation in four full-scale emergency exercises for nuclear power plants, as well as participation in several multi-agency exercises. Annex F-2 provides a list of the non-reactor exercises in which NRC participated in 2008–2010. The previous U.S. National Report discusses in more detail the emergency measures, requirements, and additional references addressing emergency response.

---

[127] See http://www.fema.gov/emergency/nrf for more detail.

Each regional office typically participates in four emergency response drills or exercises each year, selected from among the list of hostile action-based drills, fuel fabrication facility exercises, or full-scale Federal Emergency Management Agency-evaluated exercises required of U.S. nuclear facilities. Additionally, NRC headquarters participates with each of the four regional offices annually in a drill or exercise. Each regional office is expected to participate in a drill or exercise with each reactor or fuel facility site in its region, once every six years. On-scene participants during these drills and exercises include NRC licensee, and state, county, and local emergency response agencies. The scope and goals of these drills and exercises may vary to some degree, but the principal NRC objective is to demonstrate readiness to recognize and evaluate the event and to support the National Response Framework, state and local authorities, and the nuclear facility, in response to the emergency.

F.5.1.4 Incident Investigation and Event Reporting

Incident investigation is a formal process conducted to help prevent accidents. NRC's Incident Investigation program provides a formal, structured, and appropriately measured NRC investigative response to significant operational events based on their safety significance. This process includes gathering and analyzing information; determining findings and conclusions, including the causes of a significant operational event; and publishing the investigation results for NRC, industry, and public review.

The types of NRC incident investigations include:

- Establishing an Incident Investigation Team, by NRC Executive Director for Operations, for events of potentially major safety significance;
- Establishing an Augmented Inspection Team, by senior NRC management, for events of lesser safety significance; and
- Establishing a Special Inspection Team, reporting directly to the appropriate regional administrator, that focuses on a specific issue at all or a group of facilities.

NRC's Incident Investigation program, outlined in Management Directive 8.3, *NRC Incident Investigation Program*, ensures the investigation of significant events is performed in a timely, objective, systematic, technically sound, and independent way by NRC. Factual information about the event and probable cause(s) must also be documented.

A senior NRC manager reporting directly to NRC Executive Director for Operations leads the Incident Investigation Team. The team is technically and administratively supported by the Office of Nuclear Security and Incident Response. Additional information on investigation was provided in the previous U.S. National Report.

Whether or not an emergency plan is called for, NRC regulations require the timely reporting of any event that may have caused or threatens to cause release of radioactive material inside or outside of a restricted area or any event involving loss of control of licensed material, that could exceed regulatory limits (events may include fires, explosions, toxic gas releases, etc.).

93

U.S. Fourth National Report-Joint Convention on the Safety of Spent Fuel Management and on the Safety of Radioactive Waste Management

F.5.1.5 Emergency Preparedness at Radioactive Materials Facilities

NRC regulations in 10 CFR Part 30, *Rules of General Applicability to Domestic Licensing of Byproduct Material*; 10 CFR Part 40, *Domestic Licensing of Source Material*; and 10 CFR Part 70, *Domestic Licensing of Special Nuclear Material*, require some fuel cycle and materials licensees to prepare emergency plans. These emergency plans are required to comply with the requirements of 10 CFR 30.32(i)(3), 10 CFR 40.31(j)(3), or 10 CFR 70.22(i)(3). Generally, the types of information to be submitted in these emergency plans include: facility description, types of accidents, classification and notification of accidents, detection of accidents, mitigation of consequences, assessment of releases, responsibilities, notification and coordination, information to be communicated, training, safe shutdown, exercises, and hazardous chemicals.

NRC performed a regulatory analysis on emergency preparedness for nuclear fuel cycle facilities and other radioactive material licensees in 1988.[128] The analysis addressed uranium mining and milling, uranium hexafluoride conversion plants, enrichment plants, fuel fabrication, spent fuel storage, new fuel storage, reprocessing and research. For byproduct material facilities such as radiopharmaceutical operations, sealed source manufacturing, depleted uranium production and waste warehousing and burial, the study concluded accidents at these types of facilities pose a very small risk to the public. Serious accidents are infrequent and would generally involve relatively small radiation doses to a few people located in limited areas.

The most potentially hazardous accident, by a large margin, was determined to be the sudden rupture of a heated multi-ton cylinder of uranium hexafluoride. The most critical injury would be from the chemical toxicity; the accompanying radiation doses would not be significant. Prevention would be the best strategy, because, in most instances, actions taken 30 minutes after accident detection would be mostly ineffective. The most effective approach to emergency response would be a simple approach consisting of:

- Identification of accidents where protective actions should be taken off site;
- Listing the licensee's responsibilities for each type of accident, including notification of local authorities (e.g., fire and police); and
- Providing sample messages for local authorities including protective action recommendations.

Specific thematic information on emergency preparedness and planning for specific waste management facility types is summarized in the previous U.S. National Report for geologic and near-surface disposal sites, uranium mills, and decommissioning.

NRC and Agreement States issued enhanced security orders[129] that require materials licensees to have a pre-arranged plan with the local law enforcement agency (LLEA) for assistance in response to an actual or attempted theft, sabotage, or diversion of International Atomic Energy Agency (IAEA) Code of Conduct Category 1 and 2 quantities of radioactive material. The LLEA response is needed for offsite coordination, in the protection of the public health and safety, to mitigate potential consequences of malevolent use of radioactive material.

---

[128] The findings for this analysis were published in NUREG-1140, A Regulatory Analysis on Emergency Preparedness for Fuel Cycle and Other Radioactive Material Licensees.
[129] http://www.nrc.gov/security/byproduct/orders.html.

F.5.2 DOE Emergency Preparedness and Management

DOE has implemented an emergency management system for all its sites and facilities. DOE Order 151.1, *Comprehensive Emergency Management System*,[130] describes DOE's emergency management system, by establishing policy; assigning roles and responsibilities; and providing the framework for development, coordination, control, and direction. This order establishes requirements for emergency planning, preparedness, response, recovery, and readiness assurance activities and describes the approach (including a graded approach) for effectively integrating these activities under a comprehensive, all-emergency concept.

Additional emergency management details are found in DOE Guide 151.1-1A, *Emergency Management Fundamentals and the Operational Emergency Base Program.*[131] This Guide provides information about the emergency management fundamentals embedded in the requirements of DOE O 151.1C, as well as acceptable methods of meeting the requirements for the Operational Emergency Base Program, ensuring all DOE facilities have effective capabilities for all-emergency preparedness and response. Additional details on emergency management and its independent oversight are in the U.S. Third National Report, Section F.5.2.[132]

F.5.3 EPA Emergency Preparedness and Response

EPA's primary responsibilities in a radiological emergency are to perform environmental monitoring and cleanup activities (designated as Emergency Support Function 10 in the National Response Framework (NRF)).[133] EPA's specific role will vary depending on the nature of the incident. Per the Nuclear/Radiological Incident Annex of the NRF, EPA is the coordinating agency for the Federal environmental response to incidents where the radioactive material involved is not licensed, owned, or operated by a Federal agency or an NRC Agreement State. This includes incidents involving foreign, unknown, or unlicensed radiological sources that have actual, potential, or perceived radiological consequences in the U.S. or its territories (most recently, EPA provided domestic environmental monitoring for the Fukushima accident). Through its RadNet monitoring system (see Section H.4), Radiological Emergency Response Team, laboratory capabilities, and other assets, EPA works with other Federal agencies, state and local governments and first responders, and international organizations to monitor, contain, and clean up any radiological materials released to the environment. As an example, EPA's monitoring capabilities have been deployed to support responses to wildfires threatening DOE installations in New Mexico and Washington. At times, EPA's extensive assets for responding to chemical emergencies will also be involved.

EPA is also responsible for supporting state and local authorities in planning for radiological emergencies. A key aspect of this planning is the development of protective action guides (PAGs) to help emergency managers and public officials make decisions about evacuation or other actions to protect the public.[134] EPA also conducts training for first responders and participates in a wide variety of exercises.

---

[130] See: https://www.directives.doe.gov/directives/current-directives/151.1-BOrder-c/view.

[131] See: https://www.directives.doe.gov/directives/current-directives/151.1-EGuide-1a/view.

[132] http://www.em.doe.gov/pdfs/3rdpercent20USpercent20Rptpercent20onpercent20SNFpercent20JC--percent20COMPLETEpercent20REPORTpercent20-percent2010percent2013percent2008.pdf.

[133] http://www.epa.gov/radiation/emergency-response-overview.html

[134] http://www.epa.gov/radiation/rert/pags.html. EPA is in the process of updating the PAG manual.

95

U.S. Fourth National Report-Joint Convention on the Safety of Spent Fuel Management and on the Safety of Radioactive Waste Management

F.6   Decommissioning Practices (Article 26)

Both NRC and DOE have active decommissioning programs as discussed in Section D.3. Their approaches are discussed in the following subsections.

F.6.1   NRC Decommissioning Approach

NRC regulates nuclear facility decontamination and decommissioning with the ultimate goal of license termination. NRC regulations assign responsibility for decommissioning licensed and unlicensed facilities to the licensee or other responsible parties. NRC evaluates the authorized party's proposed decommissioning plan, including the licensee's justification for using a particular remediation methodology, to determine if it is appropriate. The decommissioning process consists of a series of integrated activities ending with license termination and site release. Decommissioning may be relatively simple and straightforward, or complex. Specific details on decommissioning NRC authorized facilities and activities are detailed in annual reports on the status of the decommissioning program. Information is provided on specific provisions such as timing, the review process, financial assurance, public participation and other programmatic considerations.[135]

F.6.1.1   Nuclear Reactor Facilities

Nuclear power reactor licensees must submit written certification to NRC within 30 days of a decision to permanently cease operations. Licensees must also submit a Post-Shutdown Decommissioning Activity Report and a License Termination Plan (LTP) within specified time constraints. An opportunity for a hearing is published and a public meeting is held near the facility before NRC approves the plan. NUREG-1700, *Revision 1, Standard Review Plan for Evaluating Nuclear Power Reactor License Termination Plans* describes the information requirements for an LTP. Licensees, as of 2007, must make provisions to facilitate decontamination of structures and equipment, to minimize the quantity of radioactive wastes and contaminated equipment, and to facilitate removal of radioactive wastes and contaminated materials when the facility is permanently decommissioned.

Reactor licensees may choose immediate dismantlement (DECON) or monitored deferred status (SAFSTOR) options. The choice of decommissioning method (SAFSTOR vs. DECON) is left entirely to the licensee. Licensees are not restricted to solely a SAFSTOR or DECON approach and can combine the SAFSTOR and DECON options. Current regulations require decommissioning be completed within 60 years. Additional time will be considered only when necessary to protect public health and safety.

Spent fuel can remain stored in the spent fuel pool or in dry cask storage facilities until a disposal option becomes available.

F.6.1.2   Materials Facilities and Activities

Material facilities decommissioning activities include maintaining regulatory oversight of complex decommissioning sites, conducting inspections, interacting with the affected

---

[135] NRC Annual Status of Decommissioning Program is available from http://www.nrc.gov/about-nrc/regulatory/decommissioning.html.

public, undertaking financial assurance reviews, and coordinating with other partner Federal agencies.

The initiating conditions for decommissioning a materials facility, as well as the major steps and timing for decommissioning, are documented in annual reports on the status of the decommissioning program, which were mentioned previously. There are occasions in which the licensee requests restricted release of the site. In those cases, where the authorized party proposes restricted release of the site, NRC first evaluates the compliance with the financial assurance and institutional control provisions of the decommissioning plan (DP), before considering the remainder of the DP.

NRC's staff review is guided by NUREG-1757, *Consolidated Decommissioning Guidance*, where NRC has consolidated its decommissioning guidance for materials sites into a more risk-informed and performance-based document.

NRC inspects the facility during decommissioning operations to ensure compliance with the DP. These inspections will normally include in process and confirmatory radiological surveys. LLW from decommissioning is disposed at a licensed LLW disposal facility (pending availability) after components and materials are dismantled and decontaminated. Other waste with sufficiently low concentrations of radionuclides, e.g., building rubble, can be disposed by alternate methods. See Section H.1.4.

F.6.1.3 <u>Decommissioning License Termination Criteria</u>

NRC's dose constraint for decommissioned facility unrestricted release is detailed in 10 CFR Part 20 Subpart E. Table F-3 provides the dose limit for decommissioning. License termination under restricted conditions (restricted release) is permissible when achieving the unrestricted levels would result in net public or environmental harm. This requires relying on institutional controls (10 CFR 20.1403 for specific provisions). However, during the period of performance of the institutional controls, the decommissioned facility would be expected to comply with the unrestricted release constraint.

F.6.2 DOE Decommissioning Approach

DOE's management approach for disposing excess facilities is described in DOE Order 413.3B, *Program and Project Management for the Acquisition of Capital Assets*,[136] with the technical approaches described in Order 430.1B, *Real Property Asset Management*.[137] Further guidance is provided in DOE Guide 430.1-4, *Decommissioning Implementation Guide*.[138] Additional Orders and Guides for decommissioning and cleanup related activities are found in the DOE Directives system.[139]

Most decommissioning projects are conducted under a variety of regulatory processes and site-specific cleanup agreements, which are legally binding and specify the process, end states, decision points, and required approvals. Regulatory requirements and independent oversight for these projects, as well as for spent fuel and radioactive waste

---

[136] See: https://www.directives.doe.gov/directives/current-directives/413.3-BOrder-b/view.
[137] See: https://www.directives.doe.gov/directives/current-directives/430.1-BOrder-bc2/view.
[138] See: https://www.directives.doe.gov/directives/current-directives/430.1-EGuide-4/view.
[139] See: https://www.directives.doe.gov/directives

management activities are described in Section E.2.3. Additional details on DOE decommissioning projects are described in the U.S. Third National Report.[140]

F.7 General Safety Requirements (Articles 4 and 11)

General safety requirements addressed in the subsections below were called out specifically in the report preparation guidance.[141]

F.7.1 Criticality Control and Residual Heat Removal

F.7.1.1 Criticality Control

The American Nuclear Society Standards Subcommittee 8 (ANS-8), Operations with Fissionable Materials Outside Reactors has developed national standards for the prevention and mitigation of criticality accidents during handling, processing, storing, and transporting special nuclear materials at fuels and material facilities. These national standards have been approved by the American Nuclear Society Committee N16 on Nuclear Criticality Safety and by the American National Standards Institute (ANSI). ANSI/ANS-8 nuclear criticality safety standards provide guidance and criteria on good practices for nuclear criticality safety generally acceptable to NRC for the prevention and mitigation of nuclear criticality accidents. NRC has incorporated recommendations from these sources into Regulatory Guide 3.71 *Nuclear Criticality Safety Standards for Fuels and Material Facilities.*[142]

Interim Staff Guidance on the safety of spent fuel management, including criticality and subcriticality safety are provided at: http://www.nrc.gov/reading-rm/doc-collections/isg/spent-fuel.html.

Criteria for criticality safety for the independent storage of spent fuel, HLW, and GTCC LLW are defined in NRC regulations in 10 CFR Part 72, *Licensing Requirements for the Independent Storage of Spent Nuclear Fuel, High-level Radioactive Waste*, and *Reactor-related Greater Than Class C Waste, Subpart F, Criteria for Nuclear Criticality Safety.* Section 72.124 establishes criteria for nuclear criticality safety, including design for criticality safety, methods of criticality control, and criticality monitoring.

F.7.1.2 Residual Heat Removal

Dry storage cask systems (for both HLW and spent fuel) are required to have reliable passive heat removal capability. NRC regulations and DOE orders require the decay heat removal for storage facilities be capable of reliable operation so the temperatures of materials used for systems, structures, and components important to safety, e.g., fuel assembly cladding material, and solidified HLW packages, remain within the allowable limits under normal, off-normal, and accident conditions.[143] Additionally, wet and dry fuel assembly transfer systems must also have adequate decay heat removal under normal, off normal, and accident conditions. Technical specifications for heat removal capability

---

[140] http://www.em.doe.gov/pdfs/3rd%20US%20Rpt%20on%20SNF%20JC--%20COMPLETE%20REPORT%20-%2010%2013%2008.pdf

[141] International Atomic Energy Agency, Guidelines Regarding the Form and Structure of National Reports: *Joint Convention on the Safety of Spent Fuel Management and on the Safety of Radioactive Waste Management* (INFCIRC/604/Rev. 1), Vienna, Austria, 19 July 2006

[142] Available at: http://www.nrc.gov/reading-rm/doc-collections/reg-guides/fuels-materials/rg/.

[143] See 10 CFR Part 72, subsections 72.122(h)(1) and (I) also 72.236(b),(f),(g), and (h).

for a storage system are proposed by the applicant or may result from the review and evaluation of submittals relating to those areas.

F.7.2 Waste Minimization

Waste minimization programs in the U.S. are mandated by law, regulations, and the President's Executive Order.[144] The Pollution Prevention Act of 1990[145] focused industry, government, and public attention on reducing the amount of pollution through cost-effective changes in production, operation, and raw materials use. Opportunities for source reduction are often not realized because existing regulations and the industrial resources required for compliance focus mainly on treatment and disposal. Source reduction, however, is fundamentally different and more desirable than waste management or pollution control.

EPA's Waste Minimization Program seeks to reduce or eliminate waste in manufacturing by promoting the concept of sustainability.[146] EPA works with industry, government agencies, and communities to voluntarily find ways to help them reduce the amount of waste they generate, particularly if the wastes contain one or more waste minimization priority chemicals.

Federal agencies, such as DOE, are subject to Executive orders mandating waste minimization and pollution prevention programs, particularly Executive Order 12780, *Federal Agency Recycling and the Council on Federal Recycling and Procurement Policy*, and Executive Order 12856, *Federal Compliance with Right-to-Know Laws and Pollution Prevention Requirements*. DOE has programs within the Office of Health, Safety and Security designed to reduce environmental releases and reduce the amount of waste eventually requiring treatment, storage, and disposal at DOE sites. Such activities include site-wide coordination, planning, reporting, employee awareness, assessments, incentives, cost-savings initiatives, recycling, and affirmative procurement programs.[147]

NRC regulations (10 CFR 20.1406) require applicants for licenses to "minimize, to the extent practicable, the generation of radioactive waste." The cost and availability of disposal of radioactive waste in the U.S. provides a strong incentive to waste generators to practice waste minimization. On a case-specific basis, NRC licensees are encouraged to manage their activities to limit the amount of radioactive waste they produce; those activities would be reviewed in any license application to ensure waste minimization and volume reduction practices are included.

Licensees as a practical matter take steps to reduce the volume of radioactive waste after it has been produced due to the cost of disposal at licensed commercial burial sites. Common means are compaction and incineration. Although a number of NRC licensees are authorized to incinerate certain LLW, most incineration is performed by a small number of commercial incinerators.

Additional information on minimization of waste throughout all stages of the nuclear fuel cycle, including disposal, is provided in Regulatory Guide 4.21, *Minimization of*

---

[144] See http://www.nepa.gov/nepa/regs/eos/eo13148.html.
[145] United States Code, Title 42, Sections 13101 and 13102.
[146] See http://www.epa.gov/wastes/hazard/wastemin/index.htm.
[147] DOE's Pollution Prevention home page is at http://www.hss.energy.gov/pp.

Contamination and Radioactive Waste Generation: Life-Cycle Planning, June 2008.[148]
This guidance provides examples of measures, which can be combined to support a
contaminant management philosophy. This philosophy includes prevention of
unintended release, early detection of potential releases, and aggressive cleanup when
releases happen.

F.7.3 Interdependencies within Spent Fuel and Waste Management

Successful management of spent fuel and radioactive waste requires careful integration
among power or research reactors, waste generators, storage facilities, treatment
facilities, disposal sites and their transportation interfaces (Articles 4(iii) and 11(iii)).
Integration is achieved through interface management, such as specified waste
acceptance criteria, so generators and disposers have a common understanding of the
waste. Acceptance requirements define the interfaces. The U.S. recognizes the
importance of this integration and manages the interfaces between various steps, e.g.,
storage, transportation, and disposal.

The U.S. Government uses a system composed of inspections, enforcement, quality
assurance, testing and record keeping, thereby ensuring interdependencies among
these steps remain relatively seamless. Manifests are used for transportation of
radioactive waste and spent fuel. Portal monitors and other monitors located at specific
check points are used to confirm the characteristics of radioactive materials as they are
transferred within a site, as well as in shipments between facilities. Disposal facility
operators use the monitoring results to review and verify the validity of assumptions
made and to update the assessments as specified in Article 15 for the period after
closure. The U.S. has regulations governing cradle-to-grave management of radioactive
waste, and waste managers are responsible for the safety of their inventories under the
terms of their licenses or safety bases.

F.7.4 National Laws/Regulations and International Criteria and Standards

The U.S. has an extensive and comprehensive set of laws and regulations for radiation
protection, meeting the intent of Article 4 and Article 11 of the Joint Convention. EPA
(Section E) is responsible for developing national standards on radiation protection. The
U.S. Government works with international organizations, e.g., the IAEA and the
International Commission on Radiological Protection (ICRP), to ensure U.S. standards
are in general harmony with recommendations from these organizations. NRC, DOE,
and EPA are involved in the process of revising and drafting IAEA Safety Standards
relating to nuclear, radiation, waste and transport safety. Because transportation is
excluded from the definitions of radioactive waste and spent fuel management (see
Article 2, items (i) and (n)), the activities supporting the revision of Safety Standard TS-
R-1 are not discussed in this report. However, the U.S. Government has a very active
role in the radiation safety standards committee, the waste safety standards committee,
the nuclear safety standards committee, and the Commission on Safety Standards.
These committees meet biannually to review and approve safety standards for
publication by the IAEA. Another example is the Code of Conduct on the Safety of
Radioactive Sources.

The U.S. believes these standards are a valuable source of guidance that a country can
use to establish or enhance its national programs. These standards, however, do not

---

[148] Regulatory Guide 4.21 is available at http://www.nrc.gov.

100

U.S. Fourth National Report-Joint Convention on the Safety of Spent Fuel Management and on the Safety of Radioactive Waste Management

prescribe the only approach to establishing strong national programs and are not binding on any country, except to the extent an individual country, acting in accordance with its national framework, incorporates all or part of them into its national law or regulations.

Several agencies are now using or allowing the use of the updated dose coefficients found in ICRP Publications 68 and 72. However, the U.S. has not adopted the annual dose limits in ICRP 60. New recommendations have been issued by the ICRP and most U.S. agencies are studying those changes before considering any revisions to current public and worker dose limits. Any change from effective dose equivalent to effective dose as the basis for human dosimetry has not yet occurred on a broad scale, although new regulations may incorporate the newer dose methods.

F.7.5 Biological, Chemical, and Other Hazards

The U.S. has major environmental laws taking into account biological, chemical, and other hazards. Operators of facilities must abide by these laws to protect workers, the public, and the environment. Laws are enforced through the implementation of EPA regulations. EPA in turn delegates some regulatory authority to states meeting the minimum Federal requirements. One such law is the Resource Conservation and Recovery Act (RCRA), which grants EPA the authority to control hazardous waste from "cradle-to-grave." This includes the generation, transportation, treatment, storage, and disposal of hazardous waste. RCRA also sets forth a framework for the management of non-hazardous solid wastes. The 1986 amendments to RCRA enabled EPA to address environmental problems resulting from underground tanks storing petroleum and other hazardous substances. RCRA focuses only on active and future facilities and does not address abandoned or historical sites covered by CERCLA or (Superfund).[149] The 1984 Federal Hazardous and Solid Waste Amendments to RCRA required phasing out land disposal of untreated hazardous waste. Some of the other mandates of this strict law include increased enforcement authority for EPA, more stringent hazardous waste management standards, and a comprehensive underground storage tank program. Impacts from chemical hazards are assessed as part of the environmental assessment process. These assessments are required prior to constructing spent fuel and radioactive waste management facilities.

F.7.6 Avoiding Undue Burden/Impacts on Future Generations

U.S. policy to manage, store, and dispose of spent fuel and radioactive waste is aimed at not placing undue burdens on future generations. Progress made toward timely decommissioning of inactive nuclear facilities and storage and permanent disposal of spent fuel and radioactive waste is a strong component of a strategy to minimize the impacts to future generations from these materials. Performance requirements on disposal sites mandate the level of isolation to ensure there are no undue burdens on future generations. The WIPP geologic repository for TRU waste is an example of the U.S. addressing the burden/impacts on future generations as national policy.

F.8 Existing Facilities (Articles 5 and 12)

Article 5 and Article 12 of the Joint Convention specify each Contracting Party must take steps to review safety of any spent fuel and radioactive waste management facility

---

[149] 42 U.S.C. 9601, et seq.

101

U.S. Fourth National Report-Joint Convention on the Safety of Spent Fuel Management and on the Safety of Radioactive Waste Management

existing at the time the Joint Convention enters into force and to ensure, if necessary, all reasonably practicable upgrades are made.

The U.S. conducts safety reviews of both commercial and governmental spent fuel and radioactive waste management facilities under its existing regulations. No additional reviews of existing facilities are required to comply with the Joint Convention because existing facilities are already subject to periodic safety reviews. The frequency and type of assessments and inspections depend on the type of facility and results of previous safety reviews.

F.9 Siting Proposed Facilities (Articles 6 and 13)

The U.S. has legal and regulatory structures described in Section E to site proposed new facilities. The process provides for evaluation of all relevant site related factors, safety impacts to workers, the public, the environment, and socio-economic impacts.

F.9.1 Assessing Environmental Impacts (NEPA Process)

The National Environmental Policy Act of 1969 (NEPA) is the basic National charter for protection of the environment. It establishes policy, sets goals, and provides means for carrying out the policy. Federal agencies have implementing regulations to integrate environmental values into their decision-making processes by considering the environmental impacts of their proposed actions and reasonable alternatives. The NEPA process provides the option to the public to attend NEPA-related hearings or public meetings and to submit comments directly to the lead agency. The previous U.S. National Report provides more specific information on NEPA.

F.9.2 Site Selection

NRC regulations prescribe site characterization activities and pre-license application reviews by NRC, as well as the application requirements for licensing and construction authorization. The regulations also provide for participation in the pre-licensing (site) review and licensing review by states, affected Tribal governments, and interested stakeholders. Information is made publicly available.

Site selection for a new spent fuel or waste management facility is embodied in the environmental assessment process (implementation of NEPA). Licensees select a site based on consideration of many factors, including geography, demography, meteorology, hydrology, seismology, and the geology characteristics of the site and the surrounding area. Nearby industrial, transportation, sensitive areas, park lands, historical sites, and military facilities are also a consideration in the selection process.

From the information supplied in response to the regulations, NRC can determine if the applicant has properly addressed environmental, socio-economic, and other site considerations, which could be deleteriously affected by the proposed operation or facility.

F.9.3 Public and Stakeholder Involvement

The U.S. recognizes the many benefits derived from public participation in its program activities, including spent fuel and radioactive waste management. Public participation is open, ongoing, two-way communication – both formal and informal – between

102

U.S. Fourth National Report-Joint Convention on the Safety of Spent Fuel Management and on the Safety of Radioactive Waste Management

government officials and stakeholders. Public participation provides a means for the government to gather the most diverse collection of opinions, perspectives, and values from the broadest spectrum of the public, enabling the government to make better, more informed decisions. Public participation benefits stakeholders by creating an opportunity to provide input and influence decisions. Information is available to members of the public about different topics, including decommissioning, spent fuel, and radioactive waste.[150]

Congress enacted the WIPP Land Withdrawal Act in October 1992, giving EPA significant new responsibilities for certifying DOE's determination of compliance at WIPP. EPA, in implementing its responsibilities, committed to conducting an open public process including interaction with all interested parties. This increased the public's understanding of EPA's role and responsibilities for the WIPP project, enabled the public to make informed decisions about the project by increasing their knowledge about radiation and its risks, and enhanced the overall decision-making process.[151]

Officials at many DOE sites have formed formal panels made up of interested citizens to advise the government on planned ongoing activities under the Federal Advisory Committee Act. Site-Specific Advisory Boards (SSABs) provide consensus advice and recommendations to DOE spent fuel and waste management activities at most locations where spent fuel and radioactive waste is stored.[152]

DOE, EPA and NRC conduct public hearings and public meetings, accept written and electronic comments on proposed actions, participate in stakeholder meetings, and provide internet sites.[153] NRC's internet website provides a full description of the agency's public information process and meeting calendar.

F.10 Facility Design and Construction (Articles 7 and 14)

Articles 7 and 14 of the Joint Convention require that spent fuel and radioactive waste management facilities are designed and constructed to limit possible radiological impacts and discharges throughout their life cycle. This is accomplished by performing reviews of the proposed operations against well-established design and construction criteria in the standards, regulations and orders. Subsequent monitoring and inspection during the construction process provides confidence that the operation will operate safely. Examples include DOE Order 420.1B, which requires all facilities to be designed for protection from natural phenomena and to facilitate safe decommissioning at end of their operating life. NRC, EPA and Agreement and Authorized States have similar provisions.

F.11 Assessing Facility Safety (Articles 8 and 15)

The Joint Convention requires that a systematic safety assessment and an environmental assessment appropriate to the hazards present at the facility are prepared to cover the entire life cycle. Updated and detailed assessments are required before operations commence. Although safety assessment is generally a stand-alone

---

[150] See http://www.nrc.gov/reading-rm/doc-collections/fact-sheets/.
[151] Additional information on the WIPP outreach program was provided in the Second U.S. National Report in Annex F-8. See http://www.epa.gov/radiation/wipp.
[152] Information on SSABs may be found at http://www.em.doe.gov/Pages/ssab.aspx.
[153] See http://www.epa.gov/radiation/index.html and http://www.nrc.gov.

process, it is also addressed as part of the NEPA process (see Section F.9.1). NRC employs a risk-informed and performance based approach to decision-making where risk insights are considered along with other factors such as engineering judgment, safety limits, redundancy, and diversity. Risk insights are gathered by asking three questions: "What can go wrong?" "How likely is it?" and "What are the consequences?" A risk assessment is a systematic method for addressing these three questions to understand likely outcomes, sensitivities, areas of importance, system interactions, and areas of uncertainty. This is applied agency-wide in the decision making process.

F.12 Facility Operation (Articles 9 and 16)

The U.S. uses results of inspection, monitoring, and testing to verify and review safety assessment assumptions (Article 16(iii)).

F.12.1 NRC Facility Safety

NRC regulations for issuing site-specific licenses for the operation of Independent Spent Fuel Storage Installations are in 10 CFR Part 72. These regulations incorporate a graded approach and require the licensee to demonstrate via safety assessment that the facility is operated safely. The safety assessment provides the basis for functional and operating limits that are conditions of the operating license. NRC regulations require licensees to update safety assessments whenever significant new information becomes available; possibly reducing a margin of safety or requiring a change to license conditions. Part 72 also requires the operator of an ISFSI to update its safety analysis every 24 months and update its decommissioning plan to reflect current inventories and conditions of the site and structures at the time of decommissioning.

NRC has regulations (10 CFR Part 61) and internally developed licensing and inspection programs governing the authorization to operate low-level radioactive waste disposal facilities. Part 61 requires the licensee to prepare a performance assessment of the disposal facility demonstrating that the performance objectives of Part 61 are met, which may be used by NRC to incorporate conditions in the license deemed necessary to protect public health and safety. The performance assessment must be updated whenever significant changes are made to the disposal facility and at time of closure.

Operations safety data are reported to or identified by NRC in event reports, inspection reports, component failure reports, industry reports, safeguard and security events, reports of defects and noncompliance (10 CFR Part 21), and reports of operation experience at foreign facilities. NRC screens operations safety data for safety significance, trends and generic implications, and the need for further regulatory action. NRC also develops, coordinates, and issues generic communications to alert industry to safety concerns and recommends the need for special inspections or event investigations.[154] Examples were included in previous U.S. National Reports.

F.12.1.1 Inspecting NRC-Licensed Commercial Facilities and Activities

NRC inspects licensed commercial nuclear power plants, research reactors, fuel cycle facilities, and radioactive materials activities and operations, including their management of radioactive waste and discharge of radioactive effluents. If an inspection identifies that a licensee is not in compliance with regulatory and/or license requirements, NRC

---

[154] See http://www.nrc.gov/reading-rm/doc-collections/gen-comm/.

informs the licensee of the problems found and provides the licensee an opportunity to address the problems. NRC conducts follow-up inspections to ensure problems are corrected.

NRC's safety oversight program is designed to limit exposures to acceptable limits and maintain them ALARA, protect the environment, and safeguard radioactive material from terrorist threats. The oversight program includes inspections and assessments of licensee and vendor activities with a focus on minimizing risk to public health and safety. NRC issues reports to document inspection findings. These inspection reports may contain enforcement actions and follow-up inspection items. NRC makes inspection reports electronically available for public review in its Agency-wide Documents Access and Management System (ADAMS) by searching for a site name or docket number or, in the case of spent fuel storage, a cask designer's name.[155]

NRC also conducts routine safety inspections of ISFSIs and of vendors and fabricators of dry cask storage systems. The inspectors examine whether licensees and vendors are performing activities in accordance with radiation safety requirements, licensing and certificate of compliance requirements, and quality assurance program commitments. Announced or unannounced NRC inspections are conducted during pre-operational testing, and periodically between one and three years afterward, to determine if the licensee is in full compliance. Licenses and the technical specifications included in Certificates of Compliance for cask designs contain additional inspection/review requirements. Inspectors follow guidance in NRC's Inspection Manual.[156]

NRC and Agreement States conduct approximately 2000 inspections of its nuclear material licensees per year. These inspections review areas such as training, radiation protection programs, patient dose records, and security, as well as radioactive waste and/or spent fuel management. Details are addressed in operational inspection manual chapters.[157]

F.12.1.2 NRC Enforcement and Civil Penalties

If licensees violate regulatory or license requirements, NRC initiates its enforcement process, based on the investigation of results from inspection, testing or other violation identification mechanisms, including allegations. Apparent violations are assessed in accordance with NRC's Enforcement Policy. The NRC Enforcement Policy is available to NRC licensees and members of the public on the NRC website.[158]

The Office of Enforcement exercises oversight of NRC enforcement programs, provides programmatic and implementation direction to regional and Headquarters offices conducting or involved in enforcement activities, and ensures regional enforcement programs are adequately carried out. NRC uses three primary enforcement actions: Notices of Violations (NOVs), Civil Penalties, and Orders.[159] Details of these actions and

---

[155] See http://www.nrc.gov/reading-rm/adams.html#web-based-adams.
[156] See http://www.nrc.gov/reading-rm/doc-collections/insp-manual/.
[157] A full list is presented at
http://www.nrc.gov/reading-rm/doc-collections/insp-manual/manual-chapter/index.html#page-content.
[158] See http://www.nrc.gov/about-nrc/regulatory/enforcement/enforce-pol.html.
[159] The term *order* within this context is distinguished from a DOE order, which is a directive and/or policy for radiation protection of the public and environment that applies to DOE sites and contractors.

NRC enforcement process can be found in the previous U.S. National Report and NRC Office of Enforcement website.[160]

NRC order issuing authority under Section 161 of the AEA extends to any area of licensed activity affecting the public health and safety. NOVs and civil penalties are issued based on the significance of the violations. Orders may be issued for violations, or in the absence of a violation, to address a public health or safety issue.

A graphical representation of the NRC graded approach for dispositioning violations is included on NRC's website.[161]Civil penalties are normally assessed for severity violations, as well as deliberate violations of the reporting requirements. Additional details on severity levels and recent enforcement actions for materials facilities are available from NRC website.[162]

All orders and Alternative Dispute Resolutions (ADRs) are made available to the public. Significant enforcement actions (including actions to individuals) are included in the Enforcement Document Collection in the Electronic Reading Room of NRC's web site.

CALENDAR YEAR 2010 NRC ENFORCEMENT SUMMARY HIGHLIGHTS

NRC issued 124 escalated enforcement actions, including:
- o 84 escalated Notices of Violation without civil penalties
- o 23 proposed civil penalties totaling $ 673700
- o 16 enforcement orders associated with ADR settlements and orders prohibiting involvement in NRC-licensed activities.
- o One order imposing civil penalties
- o Three demands for information
- o The ADR process was used successfully in 10 enforcement cases

---

F.12.2 DOE Facility Operation

DOE exercises regulatory authority over spent fuel and radioactive waste management operations conducted by DOE or on its behalf pursuant to the Atomic Energy Act (see Table E-1), except in cases where Congress has specifically provided NRC authority over DOE facilities or activities (see Section E.2.3). The major applicable Federal regulations include 10 CFR 820 *Procedural Rules for DOE Nuclear Activities,*[163] 10 CFR 830 *Nuclear Safety Management,*[164] 10 CFR 835 *Occupational Radiation Protection*[165] and 10 CFR 851, *Worker Safety and Health Program.*[166] These DOE's nuclear safety Federal regulations are similar to NRC's.

Facility operations and radiation protection programs fall under standards and requirements established in DOE Orders and Directives. The major applicable orders include DOE Order 458.1, *Radiation Protection of the Public and the Environment*[167] and DOE Order 435.1, *Radioactive Waste Management.*[168]

---

[160] See http://www.nrc.gov/about-nrc/regulatory/enforcement/program-overview.html.
[161] Ibid.
[162] See http://www.nrc.gov/reading-rm/doc-collections/enforcement/actions/materials/.
[163] See: http://ecfr.gpoaccess.gov/cgi/t/text/text-idx?c=ecfr&sid=878197d37c47d5d267017ca6fde249ba&rgn=div5&view=text&node=10:4.0.2.5.24&idno=10
[164] See: http://ecfr.gpoaccess.gov/cgi/t/text/text-idx?c=ecfr&tpl=/ecfrbrowse/Title10/10cfr830_main_02.tpl
[165] See: http://ecfr.gpoaccess.gov/cgi/t/text/text-idx?c=ecfr&tpl=/ecfrbrowse/Title10/10cfr835_main_02.tpl
[166] See: http://ecfr.gpoaccess.gov/cgi/t/text/text-idx?c=ecfr&tpl=/ecfrbrowse/Title10/10cfr851_main_02.tpl
[167] See: https://www.directives.doe.gov/directives/archive-directives/458.1-BOrder-admc1
[168] See: https://www.directives.doe.gov/directives/current-directives/435.1-BOrder-c1/view

Other requirements implementing 10 CFR Part 830 are found in DOE Order 422.1, *Conduct of Operations.*[169] Implementation guidance is found in DOE G 421.1-2, *Implementation Guide For Use in Developing Documented Safety Analyses To Meet Subpart B of 10 CFR 830,*[170] and DOE G 423.1-1A, *Implementation Guide For Use In Developing Technical Safety Requirements.*[171] Table E-2 provides a list of spent fuel and radioactive waste management Federal regulations and DOE Orders and other Directives. Further details are found in the U.S Third National Report, Section F.12.3.

Oversight responsibility for DOE facility safety is assigned to the Office of Health, Safety and Security (HSS). In carrying out this responsibility, HSS is focused on providing effective and consistent safety-related policy development, technical assistance, education and training, and complex-wide independent oversight and enforcement. Key safety functions include:

- **Corporate Safety Analysis** providing analysis and certification of DOE-wide performance in protecting the public, the workers, and the environment while performing the missions of DOE. This analysis supports corporate decision-making and synthesizes operational information to support continuous environment, safety and health improvement across the DOE complex. Such analysis is a means of communicating experiences to potentially reduce risk, improve efficiency, and enhance the cost-effectiveness of DOE processes and operations.
- **Corporate Safety Programs** including safety program topics such as Accident Investigation, Accident/Incident Reporting System, Analytical Services Program, Behavior-Based Safety/Human Performance, Corrective Action Management Program, Federal Occupational Safety and Health, Laboratory Accreditation Program.
- **Worker Safety and Health Enforcement** implementing DOE's congressionally mandated requirements for a worker safety and health program that reduces or prevents occupational injuries, illnesses, and accidental losses by providing DOE contractors and their workers with safe and healthful workplaces at DOE sites; and procedures for investigating whether a violation of a requirement of this regulation has occurred, determining the nature and extent of any such violation, and imposing an appropriate remedy.

Detailed information about the HSS mission, functions, and programs may be found at http://www.hss.doe.gov/whoweare.html

---

[169] See: https://www.directives.doe.gov/directives/current-directives/422.1-BOrder/view
[170] See: https://www.directives.doe.gov/directives/current-directives/421.1-EGuide-2/view
[171] See: https://www.directives.doe.gov/directives/current-directives/423.1-EGuide-1/view

107

U.S. Fourth National Report-Joint Convention on the Safety of Spent Fuel Management and on the Safety of Radioactive Waste Management

G. SAFETY OF SPENT FUEL MANAGEMENT

Section F described aspects common to spent fuel and radioactive waste safety per Articles 4-9 of the Joint Convention. This section provides additional information relative to the same Articles pertaining solely to spent fuel. This section also addresses Article 10 of the Joint Convention.

G.1 General Safety Requirements (Article 4)

The need for general safety requirements is found in the Atomic Energy Act (AEA) and the Nuclear Waste Policy Act, as amended. The licensing requirements for storage of spent fuel, high-level waste (HLW), and reactor related Greater-than-Class C LLW waste (GTCC) at an independent spent fuel storage installation are contained in 10 Code of Federal Regulations (CFR) Part 72. Additional applicable regulations include 10 CFR Part 71, *Packaging and Transportation of Radioactive Material*; Part 73, *Physical Protection of Plants and Materials*; Part 75, *Safeguards on Nuclear Material-Implementation of US/IAEA Agreement*. Table E-2 lists key NRC regulations.

Although both pool storage and dry storage are safe methods for spent fuel management, there are significant differences. Pool storage requires greater and more consistent operational vigilance by utilities or other licensees and the satisfactory performance of many mechanical systems using pumps, piping and instrumentation, whereas dry cask storage systems rely on passive measures to ensure safety.

Nuclear Regulatory Commission (NRC) authorizes storage of spent fuel at independent spent fuel storage installations (ISFSIs) under two licensing options: site-specific and general licenses. To obtain a site-specific license, an applicant submits an application to NRC and NRC performs a technical review of all aspects of the proposed ISFSI. The application must contain general and financial information; the applicant's technical qualifications to be able to safely operate the ISFSI; a safety analysis report; quality assurance program; an operator training program; physical protection, decommissioning, and emergency plans; an environmental report; and proposed license conditions. Upon approval, NRC issues a license for a 40–year term. NRC recently completed a rulemaking to extend the license and certificate of compliance from 20-year to 40-year terms. The licensee has an option for renewal at the end of the license term.[172]

A general license to store spent fuel at an ISFSI is automatically granted, via 10 CFR 72.210, to any nuclear power plant licensee that has a license to either operate or possess nuclear power reactors under 10 CFR Part 50 or 10 CFR Part 52. The general license is valid for 40 years from the loading date of each storage cask, as long as the as the licensee maintains its 10 CFR Part 50 or 10 CFR Part 52 license and continues to meet the other requirements of the general license.

The prospective general licensee must also review its security program, emergency plan, quality assurance program, training program and radiation protection program, and make any necessary changes to incorporate the ISFSI at its reactor site.

---

[172] More specific information about the licensing process for both wet and dry storage facilities can be found at http://www.nrc.gov/waste/spent-fuel-storage.html.

109

U.S. Fourth National Report-Joint Convention on the Safety of Spent Fuel Management and on the Safety of Radioactive Waste Management

NRC has approved 16 dry storage systems, which potential general licensees may consider for use at their site. These designs are listed in NRC regulations (10 CFR 72.214). An NRC-approved storage cask has been technically reviewed for its safety aspects and found adequate to store spent fuel because it meets NRC's requirements in 10 CFR Part 72. The Certificate of Compliance (CoC) expires 40 years from the date of issuance with a renewal option.

NRC approves dry cask storage systems by evaluating each design for resistance to normal and off-normal conditions of use and hypothetical accident conditions such as floods, earthquakes, tornados, and temperature extremes. The maximum allowable heat generation from the fuel assemblies stored in each cask may be different for each design. The temperature of the fuel in the casks continuously decreases over time. There have been no releases of spent fuel storage cask contents or other significant safety problems from the dry cask storage systems in use today.

G.2 Existing Facilities (Article 5)

ISFSIs in the U.S. use different dry storage system designs (some of which are licensed for a specific site). The designs encompass dual-purpose canisters, vault storage systems, and metal and concrete storage casks. These storage casks are made by several vendors and have been licensed or certified by NRC. Almost all ISFSIs are owned and operated by 10 CFR Part 50 power reactor license holders.

Typical examinations for renewal of a storage license or CoC evaluate aging of components through corrosion, chemical attack, and other mechanisms for reduction in the efficacy of important storage cask components. Current guidance on renewing site-specific storage licenses or CoCs is contained in NUREG-1927, *Standard Review Plan for Renewal of Spent Fuel Dry Cask Storage System Licenses and Certificates of Compliance.*[173]

G.3 Siting Proposed Facilities (Article 6)

Siting of ISFSIs at operating nuclear power plants is addressed within the context of the safety case associated with the operating facility. For the case of siting ISFSIs at an away-from-reactor facility, 10 CFR Part 72 Subpart E, *Siting Evaluation Factors*, addresses factors such as the radiological criteria, design basis events, geologic considerations, and controlled areas.[174]

A private initiative, Private Fuel Storage, LLC (PFS), submitted an application in June 1997 to NRC to construct an ISFSI designed to accept fuel from multiple utilities. This was only the second application submitted to NRC for an away-from-reactor ISFSI and the first for storage of spent fuel from more than one utility. PFS is a consortium of nuclear utilities proposing to lease land for the proposed ISFSI from the Skull Valley Band of Goshute Indians. The PFS application sought approval for the storage of a maximum of 40,000 metric tons uranium of spent fuel at the site. The Skull Valley Band's tribal land is located southwest of Salt Lake City, Utah.

---

[173] Publicly available from NRC, http://www.nrc.gov/reading-rm/adams.html – ADAMS Accession Number ML111020115.

[174] For more detailed information on spent fuel storage, see http://www.nrc.gov/waste/spent-fuel-storage.html.

NRC issued a license to PFS on February 21, 2006, but conditioned construction authorization on the company first arranging for adequate funding. On February 21, 2007, progress in developing the facility was indefinitely delayed by actions of the U.S. Department of the Interior, which disapproved the lease arrangement between PFS and the Skull Valley Band and denied PFS the use of public lands for an intermodal transfer facility. The 10th Circuit Court of Appeals vacated decisions by the U.S. Department of Interior that blocked construction of PFS in June 2010. The ruling returned PFS' application for a right-of-way and lease of tribal land to the Department of Interior for further consideration. PFS' request is still being considered by the Department of Interior.

G.4   Spent Fuel Management Facility Design and Construction (Article 7)

General design criteria contained in 10 CFR Part 72 Subpart F establish the design, fabrication, construction, testing, maintenance, and performance requirements for structures, systems, and components important to safety, as defined at 10 CFR 72.3. These are minimum requirements for the design criteria for an ISFSI or monitored retrievable storage installation.

Subpart L in 10 CFR Part 72 establishes requirements for spent fuel storage cask design approval and fabrication for use by general licensees. This subpart also contains requirements and conditions for renewal of designs having an NRC CoC; record keeping and reporting requirements; procedures for amending a CoC; and periodic updating of safety analysis reports. Quality assurance requirements, which apply to both the facility and certificate holder, are located in 10 CFR Part 72, Subpart G.

NRC reviews safety analysis reports using guidance to the staff in NUREG-1536, *Standard Review Plan for Dry Cask Storage Systems*, and NUREG-1567, *Standard Review Plan for Spent Fuel Dry Storage Facilities*. These plans ensure the quality and uniformity of NRC reviews.

G.5   Assessing Facility Safety (Article 8)

Technical evaluations of ISFSI safety are performed in six major areas: (1) siting; (2) operating systems; (3) criteria and technical design; (4) radiation safety programs supporting protection of both worker and public health and safety; (5) accidents; and (6) proposed technical specifications. Additional details and specific requirements are contained in NUREG-1567, *Standard Review Plan for Spent Fuel Dry Storage Facilities*.[175]

Demonstrating compliance with long-term performance requirements, by necessity, will involve the use of complex predictive models supported by data from field and laboratory tests, site-specific monitoring, and natural analog studies supplemented with prevalent expert judgment.

G.6   Facility Operation (Article 9)

NRC applies its regulations, licensing and inspection programs to authorize storage of spent fuel or reactor related GTCC LLW at an ISFSI; approve the storage cask design; and ensure safe operation of the ISFSI. There have been no releases of spent fuel

---

[175] NUREG-1567 can be accessed at http://www.nrc.gov/reading-rm/doc-collections/nuregs/staff/.

111

U.S. Fourth National Report-Joint Convention on the Safety of Spent Fuel Management and on the Safety of Radioactive Waste Management

storage cask contents or other significant safety problems from the dry cask storage systems in use today.

Inspections ensure safe operation and continued integrity of the fuel in the storage cask. Other than review of indirect parameters there are no periodic inspections required to determine damage to the contents as part of the facility license conditions. It is important to note that studies have been performed on dry storage system canisters and their contents, which show no degradation that would warrant changing the storage systems licensing bases. The results of these studies are located in NUREG/CR-6381, *Examination of Spent PWR Fuel Rods after 15 Years in Dry Storage*.[176]

NRC issued guidance on the standard format and content of technical specifications and recommendations on the most important fuel parameters in NUREG-1745, *Standard Format and Content for Technical Specifications* for 10 CFR Part 72 Cask Certificates of Compliance, and NUREG/CR-6716, *Recommendations on Fuel Parameters for Standard Technical Specifications for Spent Fuel Storage Casks*.[177] The important parameters are those with a large influence on criticality safety and radiation shielding doses. The ultimate determination of parameters is based on those the applicant uses in its modeling to demonstrate safety of the storage cask and facility design.

Requirements for incident reporting are located in 10 CFR 72.74, §72.75, and §72.80. The rules require reporting significant events where NRC may need to act to maintain or improve safety or to respond to public concerns. All events are considered against the International Nuclear Event Scale (INES). A report is generated per INES requirements if the event is classified a Level 2 or above. Section F.12 provides additional information on facility operations.

G.7 Spent Fuel Disposal (Article 10)

Spent fuel is being safely stored. The Blue Ribbon Commission on America's Nuclear Future is evaluating alternative approaches concerning the back end of the fuel cycle and developing recommendations for management and disposal of spent fuel and HLW. See Section A.5.1.

---

[176] Publicly available from NRC, http://www.nrc.gov/reading-rm/adams.html – ADAMS Accession Number ML032731021.
[177] Publicly available from NRC, http://www.nrc.gov/reading-rm/adams.html – ADAMS Accession Numbers ML011940387 and ML010820352; respectively.

H. SAFETY OF RADIOACTIVE WASTE MANAGEMENT

Section F described common elements of spent fuel and radioactive waste safety per Articles 11-16 of the Joint Convention. This section provides additional information for the same Articles pertaining only to radioactive waste management. This section also addresses Article 17 of the Joint Convention.

The primary legal basis and agency responsibilities for management of radioactive waste are discussed in detail in Section E. The Nuclear Regulatory Commission (NRC) regulates commercial radioactive waste including high-level waste (HLW), low-level waste (LLW), and uranium mill tailings. NRC's regulatory framework for disposing and managing commercial spent fuel is described in Sections F and G. This section addresses NRC's safety requirements for LLW and uranium recovery programs. See Section B.2.3.2 for additional information on waste types.

DOE's waste management practices are described in DOE Order 435.1, *Radioactive Waste Management*. This order and its implementing manual require that all DOE radioactive waste be managed to protect worker and public health and safety, and the environment. DOE Order 435.1 applies to all DOE radioactive waste classes, including HLW, transuranic (TRU) waste, and LLW. The requirements span the life cycle of waste management facilities from planning through decommissioning and closure.

H.1 Existing Commercial LLW Management Facilities and Past Practices (Article 12)

H.1.1 Currently-Licensed LLW Facilities

The commercial sector's LLW is typically stored on site by licensees or by third party waste processors, either until it has decayed away (can be disposed of as ordinary trash) or until amounts are large enough for shipment to a LLW disposal site, or in the case of most disused and unwanted sealed sources, until a disposition pathway is identified.[178] LLW disposal occurs at commercially operated LLW disposal facilities and could be licensed by either NRC pursuant to 10 Code of Federal Regulations (CFR) Part 61 or Agreement States pursuant to their regulations that are compatible with 10 CFR Part 61.[179] At the present time, all sites are licensed by Agreement States. Facilities must be designed, constructed, operated, and closed to meet rigorous safety standards. The operator of the facility must also extensively characterize the facility site and analyze how the facility will be protective of public health, safety and the environment for thousands of years.

NRC regulations (10 CFR 61.12(k)) for land disposal of LLW require license applicants to submit a description of their radiation safety program for control and monitoring of radioactive effluents to ensure compliance with the radiation dose limits for the general population. Also, as required by 10 CFR 61.53(c), the licensee must maintain a monitoring program during the construction and operation of a LLW disposal facility. The monitoring system must be capable of providing early warning of releases of

---

[178] For additional information on LLW, see NUREG/BR-0216, *Radioactive Waste: Production, Storage, Disposal, Revision 2* and NRC's fact sheet on "Low-Level Radioactive Waste."
[179] There are no disposal facilities currently licensed by NRC for disposal of GTCC LLW. GTCC LLW is stored until a disposal facility is established in accordance with the LLRWPAA.

113

radionuclides from the disposal site before they leave the site boundary. Similarly, per 10 CFR 61.53(d), the licensee responsible for post-operational surveillance of the disposal site must maintain a monitoring system capable of providing early warning of releases after the site is closed.

The Low-Level Radioactive Waste Policy Act of 1980 (LLRWPA) gave states responsibility for providing disposal capacity for LLW generated within their borders (except for certain waste generated by the Federal government). The LLRWPA authorized states to enter into compacts allowing them to dispose of waste at a common disposal facility and exclude waste from states outside the compact. Most states have entered into compacts. The LLRWPA was amended in 1985 to add a series of milestones and penalties in order to encourage compliance. Figure H-1 shows the makeup of U.S. regional compacts for LLW disposal. There are now 10 compacts, comprising 42 states, and 10 unaffiliated states. The District of Columbia and Puerto Rico are considered states by the AEA and LLRWPA. Existing U.S. commercial LLW disposal sites are discussed in Section D.2.2.2. All are in Agreement States.

H.1.2 Past Practices and Formerly Licensed Facilities

Because of concerns about the criteria and procedures used for the decommissioning of sites for which licenses had been terminated, NRC reviewed such sites to assure previously licensed facilities were properly decontaminated and posed no threat to public health and safety. The Oak Ridge National Laboratory (ORNL) then reviewed all terminated materials licenses to identify sites with potential for meaningful residual contamination, based on information in the license documentation, and to identify sealed sources with incomplete or no accounting, thus representing a public hazard. ORNL examined more than 37000 terminated license files. ORNL identified about 675 material licenses and 565 sealed source licenses requiring further review. NRC either performed a follow-up review or transferred responsibility for the follow-up review to the appropriate Agreement State.

Thirty-nine formerly licensed sites were found to have residual contamination levels exceeding NRC's criteria for unrestricted release. These sites were listed in the U.S. National Report for the Second Review Meeting.[180] These sites are still in the process of decommissioning, under Regional review, or have been transferred to an Agreement State or other Federal agency. Some of the formerly licensed sites are being or have been addressed as part of the decommissioning of complex material sites. See Annex D-6.

H.1.3 Management Strategies for Low Activity Waste Sites

Management and disposal of "low-activity waste" (LAW), for example, slightly contaminated solid materials, is receiving increased attention both internationally and domestically. Although the U.S. has no official legal definition for LAW, it is a term frequently used by other nations and organizations involved in radioactive waste management.[181] One of the primary reasons LAW has become a focus of attention is the unusually large volumes to be managed in comparison to conventional LLW from the

---

[180] See http://www.em.doe.gov/pdfs/Second_National_Report--Final_Rev_30.pdf.

[181] IAEA Symposium on Low-Activity Radioactive Waste Disposal; Cordoba, Spain. December 13-17, 2004.

114

U.S. Fourth National Report-Joint Convention on the Safety of Spent Fuel Management and on the Safety of Radioactive Waste Management

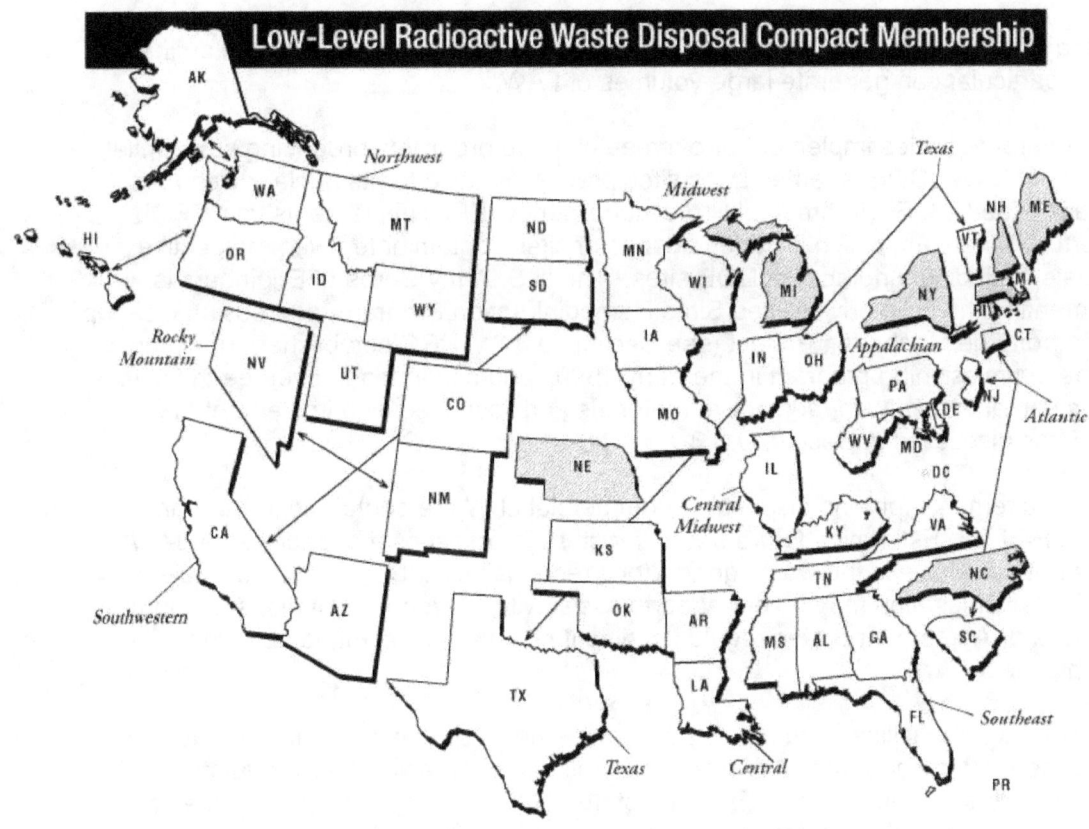

Low-Level Radioactive Waste Disposal Compact Membership

Appalachian Compact
Delaware
Maryland
Pennsylvania
West Virginia

Atlantic Compact
Connecticut
New Jersey
South Carolina

Central Compact
Arkansas
Kansas
Louisiana
Oklahoma

Central Midwest Compact
Illinois
Kentucky

Northwest Compact
Alaska
Hawaii
Idaho
Montana
Oregon
Utah
Washington
Wyoming

Midwest Compact
Indiana
Iowa
Minnesota
Missouri
Ohio
Wisconsin

Rocky Mountain Compact
Colorado
Nevada
New Mexico

Northwest accepts Rocky
Mountain waste as agreed
between compacts

Southeast Compact
Alabama
Florida
Georgia
Mississippi
Tennessee
Virginia

Southwestern Compact
Arizona
California
North Dakota
South Dakota

Texas Compact
Texas
Vermont

Unaffiliated States
District of Columbia
Maine
Massachusetts
Michigan
Nebraska
New Hampshire
New York
North Carolina
Puerto Rico
Rhode Island

Unaffiliated states are shaded

Figure H-1 U.S. Low-Level Waste Compacts[182]

---

[182] See http://www.nrc.gov/waste/llw-disposal/licensing/compacts.html

115

U.S. Fourth National Report-Joint Convention on the Safety of Spent Fuel Management and on the Safety of Radioactive Waste Management

ongoing operations of nuclear facilities. Decommissioning or cleanup of contaminated sites in particular can generate large volumes of LAW.

Four Federal agencies implement or oversee cleanup programs producing substantial amounts of LAW. DOE is remediating sites previously used for its nuclear weapons program. The U.S. Environmental Protection Agency (EPA) implements its CERCLA (Superfund) program, which includes dozens of sites contaminated with radioactive materials, including a number of DOE sites. The U.S. Army Corps of Engineers is implementing the Formerly Utilized Sites Remedial Action Program, addressing cleanup of sites from the Manhattan Project (see Section D.3.2). NRC established the complex sites decommissioning program in the early 1990s and continues to oversee the cleanup of sites contaminated with radioactive materials and licensees who implement the cleanup as discussed in Section D.3.3.

Siting requirements for land disposal of commercial LLW are contained in Subpart D of 10 CFR Part 61. Estimating future needs for storage or disposal capacities for LAW, LLW, or mixed waste is the waste generator's responsibility. DOE/National Nuclear Security Administration may be engaged to assist with the removal of disused sources from NRC or Agreement States regulated sites if no commercial disposal pathway is available.

Hazardous waste facilities and municipal or industrial solid waste landfills are now used by U.S. generators for some LAW disposal. Both types of facilities are regulated under RCRA, which is implemented by EPA and states authorized by EPA in the case of hazardous waste, and by states alone in the case of non-hazardous solid waste. Neither type of facility was originally designated for radioactive wastes. The same containment and isolation technology used in the design for hazardous and municipal solid waste is relied upon, in certain cases, for radioactive waste. See Section E.2.2.3 on mixed waste regulation.

Licensees in non-Agreement States are required to get NRC approval for disposal of such very low activity waste in RCRA cells (10 CFR 20.2002). The approval request must identify amounts, concentrations, and specific radionuclides and a performance assessment demonstrating exposures will be no more than a few tens of µSv (few millirem). NRC approval exempts such waste from further regulation with regard to its radioactive content.

Agreement States have regulations similar to 10 CFR 20.2002. NRC provides guidance direction in the form of letters to help Agreement States evaluate such 10 CFR 20.2002 disposal requests. Some examples are STP-03-003 on *Controlled Release of Concrete*, and STP-01-081, *Case-Specific Licensing Decisions on Release of Soils from Licensed Facilities*.[183]

A number of DOE sites, on a case-by-case basis, in coordination with state regulators, have limited approval for waste disposal at specific solid waste landfills. The authorized limits are established to ensure no special regulatory requirements beyond those already in place for the landfill are necessary.

LAW from remediation of sites and decommissioning is also affected by risk management decisions for the release of sites. LAW from contaminated sites may be

---

[183] See http://nrc-stp.ornl.gov/asletters/.

allowed to remain onsite under certain circumstances, often after the more highly radioactive materials have been removed. DOE plans to leave residual radioactivity in place at many sites, and will require long-term management (institutional controls) to ensure future use of the land is safe and barriers are functioning as intended. Several DOE sites have waste disposal onsite in CERCLA disposal cells requiring long-term stewardship. See Section D.2.2.2.

The Superfund program administered by EPA has a long history of permitting residual materials, both chemicals and radioactive materials, to remain on site provided a reliable system of institutional controls is established. CERCLA requires a review every five years to ensure the controls are continuing to function.

H.1.4 Controlling Solid Materials Disposition

Currently, NRC generally addresses the release of solid material on a case-by-case basis using license conditions and existing regulatory guidance. In each case, material may be released from a licensed operation with the understanding and specific acknowledgment that the material may contain very low amounts of radioactivity, but the concentration of radioactive material is so small that its control through licensing is no longer necessary. Some materials are only contaminated on or near the surface; others are contaminated throughout their volume. The regulatory processes for release of these materials are different.

Land disposal is another option for disposition of low activity radioactive material from licensed facilities (Section H.1.3). NRC can consider specific licensing actions, as well as generic requests, concerning the disposition of solid materials. NRC regulations do not contain generally applicable standards for the disposition of solid materials with relatively small amounts of radioactivity in, or on, materials and equipment. The offsite disposition of solid materials prior to license termination will continue to be evaluated on a case-by-case basis using existing guidance.

H.1.4.1 Surface Contaminated Radioactive Material Release

Criteria used by licensees to determine whether the material may be released are approved during the initial licensing or license renewal of a facility, as part of the facility's license conditions or radiation safety program. The licensees' actions must be consistent with the requirements of 10 CFR Part 20 (e.g., 10 CFR 20.1501). Thus, the licensee performs a survey of the material prior to its release. However, there are differences in the way reactor licensees and materials licensees apply the criteria for release of this material.

Nuclear reactor licensees historically follow a policy that was established and documented in NRC Office of Inspection and Enforcement Circular 81-07 and Information Notice 85-92.[184] The reactor licensees survey equipment and material before its release to identify the presence of controlled radioactive material above

---

[184] These documents are available from NRC's website: Information Notice No. 85-92: *Surveys of Wastes Before Disposal From Nuclear Reactor Facilities* (December 2, 1985) http://www.nrc.gov/reading-rm/doc-collections/gen-comm/info-notices/1985/in85092.html, and IE Circular No. 81-07: *Control Of Radioactively Contaminated Material* (May 14, 1981) http://www.nrc.gov/reading-rm/doc-collections/gen-comm/circulars/1981/cr81007.html.

natural background levels; if this "presence" is detected, then no release may occur.[185] Otherwise, the solid material in question can be released for unrestricted use.

For materials licensees, NRC usually authorizes the release of solid material through specific license conditions. One set of criteria that is used to evaluate solid materials before they are released is contained in Regulatory Guide 1.86, *Termination of Operating Licenses for Nuclear Reactors.*[186] A similar guidance document is Fuel Cycle Policy and Guidance Directive FC 83-23, *Guidelines for Decontamination of Facilities and Equipment Prior to Release for Unrestricted Use or Termination of Byproduct, Source or Special Nuclear Materials Licenses.* Both documents contain a table of surface contamination criteria which may be applied by licensees for use in demonstrating that solid material with surface contamination can be safely released with no further regulatory control. These surface contamination criteria have been used in many contexts for all types of licensees for many years.

H.1.4.2 Volumetrically Contaminated Radioactive Material Release

NRC has not provided generic guidance for unrestricted release of volumetrically contaminated materials. These situations are addressed on an individual basis, typically seeking to assure, by an evaluation of doses associated with the proposed release of the material, that maximum doses are a small percentage of the 10 CFR Part 20 dose limit for members of the public (i.e., 1 mSv/a). The practice over the years has been to allow the release of material with slight levels of volumetric contamination based on a case-by-case evaluation. These evaluations follow guidance discussed in a number of NRC documents.[187]

Reactor facilities release volumetrically contaminated materials under the provisions of Information Notice No. 88-22, *Disposal of Sludge from Onsite Sewage Treatment Facilities at Nuclear Power Stations.*[188] Certain materials may be surveyed using a representative sample and gamma spectrometry analytical methods. The provision requires that materials can be released if no licensed radioactive material above natural background levels is detected, provided the radiation survey used a detection level that is consistent with the lower limit of detection (LLD) values used to evaluate environmental samples. There are several different acceptable survey applications of the environmental LLDs and applications have included a variety of environmental media including soils, sediments, liquids and slurries.[189]

Materials licensees release volumetrically contaminated materials under the provisions of the December 27, 2002, *NRC Memorandum Update on Case-Specific Licensing Decisions on Controlled Release of Concrete from Licensed Facilities.*[190] (This memorandum indicates that controlled releases of volumetrically contaminated concrete

---

[185] The fact that no radioactive material above background is detected does not mean that none is present; there are limitations on detection capability. In practice, the actual detection capability of survey instruments is typically consistent with the criteria in NRC Regulatory Guide 1.86.

[186] Available at http://pbadupws.nrc.gov/docs/ML0037/ML003740243.pdf.

[187] NRC All-Agreement States letters (STP-00-070 [http://nrc-stp.ornl.gov/asletters/program/sp00070.pdf dated August 22, 2000], STP-01-081 [http://nrc-stp.ornl.gov/asletters/program/sp01081.pdf dated November 28, 2001], STP-03-003 [http://nrc-stp.ornl.gov/asletters/program/sp03003.pdf dated January 15, 2003]).

[188] See http://www.nrc.gov/reading-rm/doc-collections/gen-comm/info-notices/1988/in88022.html.

[189] See NRC Regulatory Guide 4.8, *Environmental Technical Specifications for Nuclear Power Plants.*

[190] See http://nrc-stp.ornl.gov/asletters/program/sp03003.pdf.

may be approved, pursuant to 10 CFR 20.2002, under an annual dose criterion of a "few mrem" (few 10s of μSv).)

H.2   DOE Waste Management Facilities

General safety requirements for DOE facilities were discussed in Section F. The following subsections contain additional information on the safety of radioactive waste management at DOE facilities. DOE manages radioactive waste owned or generated by DOE, including waste from atomic energy defense activities and waste resulting from DOE cleanup activities. Requirements for waste management are provided in DOE Order 435.1, *Radioactive Waste Management*.

H.2.1 Past Practices (Article 12)

Some past radioactive waste management practices require additional environmental restoration activities or interventions as new technology and additional characterization information become available. Examples of these past practices include waste disposal techniques, such as soil columns or crib trenches, as well as sites where remaining residual radioactivity does not meet today's standards for unrestricted release. Environmental restoration activities resulting in off-site management and disposal of radioactive waste must meet the applicable requirements of DOE Order 435.1. Cleanup decisions are reached through a formal regulatory process under CERCLA or RCRA regulations (for mixed waste).

H.2.2 Siting Proposed Facilities (Article 13)

In addition to the requirements in DOE Order 435.1, DOE radioactive waste management facilities, operations, and activities are designed and sited in accordance with DOE Order 420.1A, *Facility Safety*, and DOE Order 430.1B, *Real Property Asset Management*. Proposed locations for radioactive waste management facilities are evaluated to identify features to be avoided or must be considered in facility design and analyses. Criteria for siting a proposed new facility or expansion of an existing facility must consider:

- Environmental and geotechnical suitability;
- Human activity prohibiting site use;
- Suitability for the volume of proposed waste disposal;
- Presence of flood plain, tectonic, or water table fluctuation characteristics; and
- Ability to control radionuclide migration pathways, surface erosion and run-off.

H.2.3 Design and Construction (Article 14)

Design and construction was generally discussed in Section F.10. Generally applicable requirements and procedures for nuclear facility design, construction and operation are in 10 CFR Part 830, *Nuclear Safety Management*; DOE Policy P 450.4, *Safety Management System Policy*; and DOE Acquisition Regulation clauses at 48 CFR 970.5223-1, 48 CFR 970.5204-2, and 48 CFR 970.1100-1. DOE's *Radioactive Waste Management Manual* requires new or modified waste management facilities subject to contamination with radioactive or other hazardous materials be designed to facilitate decontamination. A proposed decommissioning method must be included in the design.

119

U.S. Fourth National Report-Joint Convention on the Safety of Spent Fuel Management and on the Safety of Radioactive Waste Management

H.2.4 Assessing Facility Safety (Article 15)

Radioactive waste facilities, operations, and activities must have a radioactive waste management basis consisting of physical and administrative controls to ensure protection of workers, the public, and the environment. Safety is ensured through specific waste management controls (waste acceptance criteria and waste certification programs), and anchored in regulatory requirements.

DOE LLW disposal facilities are sited, designed, operated, maintained, and closed so there is a reasonable expectation to comply with the performance objectives for DOE LLW disposed of after September 26, 1988; see Table F-3.

Site-specific radiological performance assessments are prepared and maintained for DOE LLW disposal facilities. The performance assessments include calculations of potential doses over a 1000-year period after closure to representative future members of the public and calculations of potential releases from the facility to provide a reasonable expectation that the performance objectives above are met for operation and closure of the facility. Longer times are not used to assess compliance with numerical standards because of the inherently large uncertainties in extrapolating such calculations over long time frames. Although the period of performance is 1000 years, sensitivity/uncertainty analysis is extended to peak dose to increase understanding of the models used and the disposal facility performance.

Analyses are performed to demonstrate compliance with the performance objectives. The assumption of average living habits and exposure conditions in representative critical groups of individuals projected to receive the highest doses is appropriate unless otherwise specified. The point of compliance corresponds to the point of highest projected dose or concentration beyond a 100-meter buffer zone surrounding the disposed waste. A larger or smaller buffer zone may be used if adequate justification is provided.

Performance assessments address reasonably foreseeable natural processes disrupting barriers against release and transport of radioactive materials. These assessments use DOE-approved dose coefficients (dose conversion factors) for internal and external exposure of reference adults, and include a sensitivity/uncertainty analysis. Performance assessments include demonstrating that projected releases of radionuclides to the environment are maintained as low as is reasonably achievable (ALARA) and include an assessment of impacts to water resources to establish limits on radionuclides disposed of near the surface.

To establish limits on the concentration of radionuclides disposed, the performance assessment also includes an assessment of impacts calculated for a hypothetical person assumed to inadvertently intrude for a temporary period into the LLW disposal facility. There is no maximum active institutional control period for DOE near-surface disposal facilities, or on the applicability of DOE's radiation protection criteria. Concerning radiation protection, the only consideration is whether risks may eventually be sufficiently low to no longer warrant continued protection. In most cases, because of the nature of the hazard and statutory requirements, DOE, or successor agencies, may be required to maintain control permanently. However, for purposes of analyzing hypothetical inadvertent intrusion, institutional controls are assumed to be effective in preventing intrusion for at least 100 years. Longer periods may be assumed with case-specific justification. The intruder analyses use performance measures for chronic and acute

120

U.S. Fourth National Report-Joint Convention on the Safety of Spent Fuel Management and on the Safety of Radioactive Waste Management

exposure scenarios, respectively, of 1 mSv (100 mrem) in a year and 5 mSv (500 mrem) total effective dose equivalent excluding radon in air.

In addition to performance assessments, site-specific radiological composite analyses are prepared and maintained for LLW disposal facilities receiving waste after September 26, 1988. The composite analysis accounts for all sources of radioactive material left at DOE sites potentially interacting with the LLW waste disposal facility, contributing to the dose projected to a hypothetical member of the public from existing or future disposal facilities. The composite analysis results are used for planning, radiation protection activities, and future use commitments to minimize the likelihood current LLW disposal activities will result in the need for future corrective or remedial actions. Additional information is in the implementation guidance for DOE Order 435.1 (DOE G 435-1).[191]

The performance assessment and composite analysis are maintained to evaluate changes affecting the performance, design, and operating bases for the facility. Performance assessments and composite analyses are reviewed and revised when significant changes alter the conclusions or the conceptual model(s). A determination of the continued adequacy of the performance assessment and composite analysis is made on an annual basis, and considers the results of data collection and analysis from research, field studies, and monitoring. Annual summaries of LLW disposal operations describe the conclusions and recommendations of the performance assessment and composite analysis and determine the need to revise the performance assessment or composite analysis.

A disposal authorization is a part of the radioactive waste management basis for a disposal facility and is obtained from DOE management prior to construction of a new LLW disposal facility. The disposal authorization statement is issued based on a review of the facility's performance assessment, composite analysis, preliminary closure plan, and preliminary monitoring plan. The disposal authorization statement specifies the limits and conditions on construction, design, operations, and closure of the LLW facility based on these reviews.

Disposal Facility Closure Plans are developed for DOE LLW disposal sites. A preliminary closure plan is developed and reviewed with the performance assessment and composite analysis. The closure plan is updated following the disposal authorization statement to incorporate conditions specified in the disposal authorization statement. Closure plans are updated as required during the operational life of the facility. Closure plans also include the total expected inventory of wastes to be disposed of at the facility over the operational life.

The final inventory of the LLW disposed in the facility is prepared and incorporated in the performance assessment and composite analysis, which is updated prior to closure. A final closure plan is prepared and implemented based on this final inventory. An updated performance assessment and composite analysis are prepared in support of the facility closure. Additional iterations of the performance assessment and composite analysis are conducted as necessary during the post-closure period.

---

[191] See http://www.directives.doe.gov/.

121

U.S. Fourth National Report-Joint Convention on the Safety of Spent Fuel Management and on the Safety of Radioactive Waste Management

H.2.5 Institutional Measures after Closure (Article 17)

Institutional control measures are integrated into land use and stewardship plans and programs, and continue until the facility can be released pursuant to DOE Order 458.1, *Radiation Protection of the Public and the Environment*. Chapter IV of DOE Order 458.1 requires, whenever property is released from radiological control, doses to the public to be below dose constraints and optimized to be ALARA. The dose constraint is 0.25 mSv/a for unrestricted release. Most radioactive waste disposal sites will not meet DOE criteria for unrestricted release at any time in the foreseeable future. The location and use of the facility is filed with the local authorities responsible for land use and zoning.

Active and passive institutional controls must be maintained until risks are sufficiently low such that continued protection would not be necessary. The active control period is determined by public risk, and some sites may indeed be released for either controlled or uncontrolled use. But in most cases, because of the nature of the hazard and statutory requirements, DOE, or successor agencies, may be required to maintain control permanently. DOE Policy 454.1 requires the maintenance of active and passive controls for as long as the hazard exists. DOE anticipates many of its facilities may never be released from active institutional control.

DOE will use active institutional controls for at least 100 years following closure. Active controls, such as fences, roadways, signs, and periodic surveillance, prevent human intrusion during this period. Ground water monitoring will continue for at least 30 years after closure, and subsidence monitoring will continue for at least 100 years. To eliminate potential overlap with hazardous waste RCRA requirements, EPA required post-closure monitoring be complementary with RCRA, so information yielded by the one monitoring program would not be duplicated by the other. In its compliance application to EPA for certification, DOE proposed these timeframes, consistent with the RCRA requirements with which it was also required to comply.

Regulations require that TRU waste disposal sites have markers and controls to inform and warn future generations about the location and purpose of this repository after the active institutional control period. These passive controls are expected to communicate the location, design, and contents of the disposal system for at least 10000 years. Planned components include: a large earthen berm, perimeter monuments, buried warning markers, magnets and metal radiation symbols, an information center using graphics and various languages, and information storage rooms. Archives will be stored in various locations around the world. A summary report is planned, and will be written in multiple languages on archival-quality paper to preserve it. While the passive institutional controls are intended to remain in effect for the entire 10000-year compliance period, it is not a "regulatory safety requirement" that they be effective for that long. The only "requirement" is to have institutional controls. EPA has allowed Waste Isolation Pilot Plant no credit toward radiation control or reduction of exposure for the long-term active institutional controls (beyond 100 years), and they have not been taken into account quantitatively.

Current planning as required for EPA certification in compliance with 10 CFR Part 194 is as follows. DOE is not required to submit final plans to EPA for institutional controls until the time of closure. For the first 100 years after closure of the repository, arrangements will be made with local law enforcement to restrict any subsurface intrusion within the withdrawn area. In addition, for the first 300 years, records will be maintained and

122

U.S. Fourth National Report-Joint Convention on the Safety of Spent Fuel Management and on the Safety of Radioactive Waste Management

maintenance will be performed on the permanent markers. After 300 years, only passive controls are currently planned at the repository site. Records will be maintained at a U.S. Government National Archives Facility beyond 300 years. The markers will be designed to last an estimated 700 years.[192]

H.3 Uranium Recovery Facilities

Uranium milling waste is designated as AEA Section 11e.(2), "byproduct material" as described in Section D.2.2.3.

H.3.1 General Safety Requirements (Article 11)

The general radiological waste safety provisions, as well as for siting and closure, for uranium milling activities are addressed in 10 CFR Part 40, with specific criteria described in Appendix A.[193] The criteria in Appendix A cover the siting and design of tailings impoundments, disposal of tailings or wastes, decommissioning of land and structures, ground water protection standards, testing of the radon emission rate from the impoundment cover, monitoring programs such as airborne effluent and off-site exposure limits, cover inspection of retention systems, financial surety requirements for decommissioning and long-term surveillance and control of the tailings impoundment, and eventual government ownership of the tailings site under a NRC general license.

A number of non-radiological constituents (e.g., ammonia, arsenic, and heavy metals) contained in tailings present a potential human health and environmental hazard to ground water and surface waters. Appendix A stipulates maximum concentration limits for ground-water protection for 14 non-radiological constituents and for radium and gross alpha. Remediation standards are made on a site-specific basis and licensees can propose as their ground-water protection standards: (1) background values; (2) maximum concentration limits per Table 5C of Appendix A; or (3) alternate concentration limits that present no significant hazard and are ALARA after considering practicable corrective actions. Compliance is assessed and assured by review of licensee monitoring and NRC inspections.

EPA regulations provide generally applicable mill standards, which NRC adopted in its regulations for uranium milling. The Office of Surface Mining, in the U.S. Department of Interior, and individual states regulate mining safety as an industrial, non-nuclear activity (i.e., uranium mining is not regulated as part of the nuclear fuel cycle). EPA also issues regulations and standards to direct the actions of other Federal agencies. NRC regulates milling and the disposal of tailings; although state agencies regulate these activities in Agreement States when the agreement specifically includes tailings. NRC requires licensees to meet EPA standards for cleanup of uranium and thorium mill sites after the milling operations have permanently closed. The annual occupational dose limit for both mines and mills is 50 mSv (5 rem). See Table F-3.

---

[192] Reference Report DOE/WIPP 04-2301, section 3.2, pages 10-15 on Records Management and section 3.4, pages 22-31 for permanent markers. In addition, reference DOE/WIPP 04-3302, *Permanent Markers Implementation Plan*.
[193] Criteria Relating to the Operation of Uranium Mills and the Disposition of Tailings or Wastes Produced by the Extraction or Concentration of Source Material from Ores Processed Primarily for Their Source Material Content.

H.3.2 Existing Facilities and Past Practices (Article 12)

The existing facilities are considered Uranium Mill Tailings Radiation Control Act (UMTRCA) Title I (closed or abandoned by 1978) or Title II (under license in 1978 or later). Reclamation at Title I facilities is largely complete with the exception of the former Atlas site near Moab, Utah; however, groundwater issues remain to be resolved at a few sites. Many Title II conventional mills in the U.S. are in decommissioning, and either have completed, or are completing, reclamation activities. The goal of the reclamation activities is to provide long-term stabilization and closure of the tailings impoundments and the sites. See Section D.2.2.3 for additional information.

NRC or the Agreement State inspects these sites at various intervals depending on the operational (or stand-by) and reclamation status. The inspection frequency can range from multiple times per year at an operating facility to once every three years at a facility in standby or reclamation status. Annex D-3A and D-3B provide status of uranium recovery facilities.

H.3.3 Siting, Design, and Construction (Articles 13 and 14)

Appendix A to 10 CFR Part 40 has 13 criteria for the siting, design, construction, operation, termination and post-closure provisions.[194] Technical Criterion 1 sets broad objectives for siting and design. The intent is to provide permanent isolation of tailings and associated contaminants by minimizing disturbance and dispersion by natural forces, and to do so without ongoing maintenance. Additional criteria specify period of performance (longevity) and other design considerations such as the presence of a liner system and dewatering method. Construction considerations include the preference for below grade disposal and reliance on a full self-sustaining vegetative cover or rock cover to reduce wind and water erosion to negligible levels.

H.3.4 Safety Assessment (Article 15)

A safety assessment is performed as part of the application review process for a uranium recovery operation. The application must provide detailed information on facilities, procedures, and equipment. The licensee must provide an environmental report with sufficient information for NRC to prepare an environmental assessment (under the provisions of the National Environmental Policy Act (NEPA) – see Table E-1) as significant changes occur during the life of the facility, e.g., expansion of the tailings pile or increasing the number of in situ recovery well fields. This information is used by NRC staff to determine that the proposed activities are protective of public health and the environment. An environmental assessment (EA) is prepared as part of this process. A more complete EIS is prepared by NRC if the EA identifies potential significant environmental impacts. The licensee may, as a result of the EIS, have to revise the design and/or increase the financial assurance mechanism, guaranteeing adequate funding for closure and disposal.

H.3.5 Institutional Measures after Closure (Article 17)

Uranium recovery licensees are required by license conditions to complete site decontamination, decommissioning, and surface and groundwater remedial actions

---

[194] These criteria can be accessed at http://www.nrc.gov/reading-rm/doc-collections/cfr/part040/part040-appa.html.

124

U.S. Fourth National Report-Joint Convention on the Safety of Spent Fuel Management and on the Safety of Radioactive Waste Management

consistent with decommissioning, reclamation, and groundwater corrective action plans before license termination.[195] Licensees must document the completion of these remedial actions in accordance with NRC procedures. This information includes a report documenting completion of tailings disposal cell construction, as well as radiation surveys and other information required by 10 CFR 40.42.

Criterion 12 of Appendix A in 10 CFR Part 40 stipulates that a final design of the waste impoundment, i.e., the final disposition of tailings, residual radioactive material, or wastes at milling sites, should assure that ongoing active maintenance is not necessary to preserve permanent isolation of the on-site waste.

Before a license is terminated, a custodial agency must accept responsibility for the long-term care and safety of the encapsulated waste. The law provides for DOE, the state, or under some circumstances, a third party, to take title of the site and to provide for any custodial activities.

The licensee will work with the custodial agency in preparing the Long Term Surveillance Plan (LTSP) because the LTSP must reflect the remediated condition of the site. The LTSP must include, among many safety-related provisions: a description of the long-term surveillance program, including proposed inspection frequency and the frequency and extent of ground water monitoring if required. Coordination between the custodial agency and the operational licensee is important and includes supplying the custodial agency with appropriate documentation such as as-built drawings of the remedial actions taken. It is the responsibility of the custodial agency to submit the LTSP to NRC for approval and acceptance. The licensee may, however, elect to help prepare the LTSP, to whatever degree is agreed upon between the licensee and the custodial agency.

When the current specific license is terminated and NRC accepts the LTSP, the custodial agency becomes the general licensee per 10 CFR 40.28(a). There is no termination of this general license. The license termination process is discussed in more detail in Section E3.0 of NUREG-1620.[196] For Title I sites, similar provisions for long-term surveillance by the custodial agency are addressed in 10 CFR Part 40, Section 40.27, as discussed in Section D.2.2.3.1.

Prior to transfer of custody, the specific licensee provides funding to cover long-term surveillance responsibilities in accordance with Criterion 10 of Appendix A. NRC will determine the final amount of this charge based on final conditions at the site.

H.4 Monitoring Releases to the Environment

RadNet, formerly known as the Environmental Radiation Ambient Monitoring System, is a national network of more than 200 monitoring stations distributed across all 50 states and the American Territories. Each station regularly samples the nation's air, precipitation, drinking water, or pasteurized milk for a variety of radionuclides (e.g., ^{131}I) and radiation types (e.g., gross beta).[197]

---

[195] See *Status of the Decommissioning Program: 2010 Annual Report*, available at: http://www.nrc.gov/about-nrc/regulatory/decommissioning.html.

[196] NUREG-1620, Revision 1, *Standard Review Plan for the Review of a Reclamation Plan for Mill Tailings Sites under Title II of the Uranium Mill Tailings Radiation Control Act*, June 2003.

[197] RadNet data and instructions for viewing reports are available through the RadNet home page at http://www.epa.gov/enviro/html/erams/index.html.

125

U.S. Fourth National Report-Joint Convention on the Safety of Spent Fuel Management and on the Safety of Radioactive Waste Management

RadNet includes:

- 124 fixed air monitors that can deliver real-time data using various telemetry methods;
- 20 older non-real-time fixed air monitors;
- 40 portable (or "deployable") real-time air monitors;
- 36 milk sampling locations;
- 31 precipitation sampling locations; and
- 73 drinking water sampling locations.

Over the past several years, EPA has been expanding the fixed network of air monitors to include near real-time capability; when completed, this capability will be available at more than 130 location nationwide. EPA's strategy for siting these new fixed station monitors is to place them in locations ensuring improved national coverage from both a population and geographic standpoint. The monitors continue to be operated by volunteers. The 40 deployable air monitors are maintained at EPA's radiation laboratories in Montgomery, Alabama, and Las Vegas, Nevada.

The expanded RadNet system is designed to provide information to help evaluate the degree and extent of contamination caused by an accidental release or a terrorist incident. The upgrades include:

- Air monitors automatically transmitting near-real-time data;
- The additional placement of monitors to improve national coverage; and
- Air monitors with the capability of being deployed after an accident or terrorist event involving radioactive materials, to allow earlier and more comprehensive coverage of the affected areas to assist in response activities and decision-making.

RadNet normally operates in a "routine" mode, sampling radiation in all media on a regularly defined schedule. RadNet operates in an "emergency" (or alert) mode in a threat of a significant radiation release, accelerating the frequency of sampling and generating many more data records for a given period of time compared to the RadNet routine mode. In response to the incident at the Fukushima Daiichi Nuclear Power Plant in March 2011, EPA increased sampling and analysis and stationed deployable monitors in several western states, as well as Guam and Saipan. In addition to real-time measurements, more than 1,700 samples were collected in the first month following the incident.

EPA also has developed RadMap, which is an interactive desktop tool to provide government at all levels with information about long-term radiation monitoring stations across the country. RadMap employs a stand-alone Geographic Information System (GIS) platform to allow users to view information about the network of more than 1700 Federal, state, local and industrial/commercial monitors. Information provided includes points of contact, available data, data collection parameters and frequency, demographics, and geographic points of interest, as well as links to the monitoring station owner's web site, if available. RadMap is a particularly useful tool for emergency responders and decision-makers to help assess the need for and appropriate location of additional monitoring resources. As a preloaded GIS program, RadMap ensures access to critical information even if computer network capabilities are unavailable.[198]

---

[198] See http://www.epa.gov/radiation/radmap/index.html.

126

U.S. Fourth National Report-Joint Convention on the Safety of Spent Fuel Management and on the Safety of Radioactive Waste Management

I. TRANSBOUNDARY MOVEMENT

I.1 U.S. Legal and Policy Framework for Transboundary Movement

The Atomic Energy Act of 1954 (AEA) assigns regulatory oversight responsibility to the Nuclear Regulatory Commission (NRC) for commercial imports and exports of source, special nuclear and byproduct materials to and from the U.S.P186F[199] P NRC regulations governing commercial imports/exports are set forth in 10 Code of Federal Regulations (CFR) Part 110. NRC amended these regulations in 1995, to conform to the guidelines of the International Atomic Energy Agency (IAEA) Code of Practice on the International Transboundary Movement of Radioactive Waste. A specific NRC license is required for imports and/or exports of radioactive materials meeting the definition of radioactive waste. They remain in force, as amended, and are consistent with the guidelines of the Joint Convention. In 2005, NRC amended Part 110 to make the regulations consistent with the current version of the IAEA Code of Conduct on the Safety and Security of Radioactive Sources, as well as the Guidance on the Import and Export of Radioactive Sources, approved by the IAEA Board of Governors and endorsed by the General Conference in September 2004. These amendments, which became effective on December 28, 2005, established specific licensing requirements for U.S. imports and exports of certain categories of radioactive sources in Part 110 Appendix P. NRC amended Part 110 again in 2010 primarily to: revise the definition of "radioactive waste" to align export and import licensing criteria with domestic regulations; clarify that for radioactive waste imports, NRC consults, as applicable, with the state where the facility is located and the low-level waste compact commission to confirm that an appropriate facility has agreed to accept and is authorized to possess the waste for management and disposal; and to remove the definition of "incidental radioactive material." The 2010 changes simplify the regulatory framework by clearly stating that exporting or importing material for the purpose of disposal or for recycling, waste treatment, or other waste management process that generates radioactive material for disposal in a 10 CFR Part 40 or Part 61 facility (or the equivalent) requires a specific export or import license from NRC.

DOE has independent authority for imports and certain exports under the AEA. Thus, DOE imports and certain exports are not subject to NRC export/import licensing regulations. For example, NRC's regulatory authority would not apply to DOE import of recovered disused sealed sources.

I.1.1 Regulatory Issues and Considerations

The U.S. began considering options to establish better domestic controls for exports and imports of radioactive wastes in the mid-to-late 1980s. There was some concern about the potential impacts of unnecessarily restricting transfers of radioactive materials which otherwise were not considered significant for nuclear weapons proliferation or as potentially endangering public health and safety if improperly handled. The process to develop and finalize U.S. regulations for the export and import of radioactive waste involved extensive review, consultation and revision.

---

[199] Although not within the scope of the Joint Convention, NRC is also responsible for imports and exports of nuclear production and utilization facilities and any equipment or components, which are especially designed or prepared for use in such facilities.

The U.S. developed the rationale for and clearly defined what additional exports and imports of nuclear materials should be controlled as radioactive waste to effectively protect public health and safety without unnecessarily curtailing international trade. It was understood that a certain amount of flexibility was needed to preserve and facilitate useful practices. The most difficult part of establishing new regulations governing the U.S. export and import of radioactive wastes was developing appropriate definitions to distinguish materials needing to be controlled from those not needing special controls.

I.1.2 Radioactive Waste

A specific license is required under 10 CFR Part 110 for imports or exports of radioactive waste, defined in 10 CFR 110.2 as any material that contains or is contaminated with source, special nuclear or byproduct materials that by its possession would require a specific radioactive material license in accordance with Chapter I of Title 10 of U.S. Code of Federal Regulations and is imported or exported for the purposes of disposal in a land disposal facility as defined in 10 CFR Part 61, a disposal area as defined in Appendix A to 10 CFR Part 40, or an equivalent facility; or recycling, waste treatment, or other waste management process that generates radioactive material for disposal in a land disposal facility as defined in 10 CFR Part 61, a disposal area as defined in Appendix A to 10 CFR Part 40, or an equivalent facility. Such radioactive waste may also contain or be contaminated with hazardous waste.[200] Radioactive waste does not include radioactive material:

- Of U.S origin and contained in a sealed source or device containing a sealed course that is being returned to a manufacturer, distributor or other entity which is authorized to receive and possess the sealed source or the device containing the sealed source;
- Containing a contaminant on any non-radioactive material (including service tools and protective clothing) used in nuclear facilities, if the service equipment is being shipped solely for recovery and beneficial reuse of the non-radioactive component in another nuclear facility and not for waste management purposes or disposal;
- Exempted from regulation by NRC or equivalent Agreement State regulations;
- Generated or used in a U.S. Government waste research and development testing program under international arrangements;
- Being returned by or for the U.S. Government or military to a facility authorized to possess the material; and
- Imported solely for the purposes of recycling and not for waste management or disposal where there is a market for the recycled material and evidence of a contract or business agreement which can be produced upon request by NRC.

The 2010 Part 110 amendment removed the definition of "incidental radioactive material." The rule incorporates aspects of incidental radioactive material into the revised definition of radioactive waste and the exclusions from the definition. The scope of the exclusion related to contamination on service equipment (including service tools) used in nuclear facilities (if the service equipment is being shipped for use in another nuclear facility and not for waste management purposes or disposal) is expanded and

---

[200] Defined in Section 1004(5) of the Solid Waste Disposal Act, 42 U.S.C. 6903(5). EPA regulates imports and exports of hazardous waste.

broadened to include some of the material that previously fell under the definition of incidental radioactive material, such as protective clothing which can be safely laundered.

I.1.3 Spent Fuel

NRC regulations adopted in 1995 only concerned LLW and did not address imports or exports of irradiated or spent fuel provisions. Exports of irradiated or spent fuel are addressed in other provisions governing exports of special nuclear material. A specific NRC license for imports of spent fuel is required if the shipment exceeds 100 kilograms. This requirement does not apply to DOE, however, because DOE has separate statutory authority to import nuclear material and equipment and is not subject to NRC import licensing.

I.2 Regulatory Requirements for Export or Import of Radioactive Waste

After an applicant seeking a license to import or export radioactive waste has provided the required information, NRC forwards the application to the U.S. Department of State (DoS). DoS is responsible for coordinating review by interested U.S. Federal Government agencies and contacting the involved foreign governments (material origin, transited, and/or destination) to either provide notice or obtain consent.

It must be determined that the proposed transaction will not be inimical to the common defense and security of the U.S. and will not result in unreasonable risks to the public health and safety before an export or import license is issued. A brief description of the process for each is provided below since the reviews for exports and imports involve different considerations.

I.2.1 Exports

For proposed exports of radioactive waste, NRC requests DoS to contact the government of the recipient nation to ask if it will accept such an import from the U.S. and requests confirmation the designated consignee is authorized to receive the radioactive waste. The DoS will ask the government to provide assurances that the material will be maintained in accordance with terms and conditions of the agreement if the material is subject to a peaceful nuclear cooperation agreement between the U.S. and the recipient nation. The U.S. accepts responses and assurances received from the nation of destination as confirmation that it has the administrative and technical capacity and regulatory structure to manage and dispose of the waste. NRC regulations do not require specific assessments and findings about the adequacy of the receiving nation's administrative and technical capacity and regulatory structure. NRC does not, however, contemplate any circumstances in which it would issue a license authorizing the export of radioactive waste to a nation without a regulated waste disposal program.

To issue a license for an export of radioactive waste, NRC must determine the action will not be inimical to the common defense and security interests of the U.S. Several factors are considered, including whether the government of the receiving nation has responded to a request from the DoS and consents to the proposed transaction. NRC will not act on an application until receiving a recommendation from interested U.S. Federal agencies including the recipient nation's consent. Additionally, export-licensing criteria are specified at 10 CFR 110.42.

Nations importing enriched uranium from the U.S. for use as reactor fuel, whether it is in the form of fresh fuel or spent fuel, must obtain U.S. consent prior to retransferring it to a third party and for reprocessing, enrichment, or other alterations in form or content under the terms and conditions of U.S. peaceful nuclear cooperation agreements. Requests for U.S. approvals of such retransfers and alterations are submitted to and processed by DOE's National Nuclear Security Administration, which coordinates U.S. interagency review of the proposed transaction. The U.S. is also consulted about the return of materials resulting from reprocessing if a nation obtains U.S. approval to transfer spent fuel to a third nation for reprocessing.

I.2.2 Imports

For proposed imports of radioactive waste, DoS contacts the government of the exporting nation and seeks acknowledgement that they are aware of the proposed transaction and any comments they might wish to provide. NRC has exclusive jurisdiction within the U.S. (the states and U.S. territories) to grant or deny specific licenses for non-DOE imports of radioactive waste. As part of the review of an application for a license to import radioactive waste, NRC consults with, as applicable, the Agreement State in which the facility is located and the low-level waste compact commission(s) to confirm that an appropriate facility has agreed to accept and is authorized to possess the waste for management or disposal. NRC will not grant an import license for waste intended for disposal unless it is clear the waste will be accepted by a disposal facility, the host state, and the compact commission (where applicable). These are among the factors considered in determining the appropriateness of the facility agreeing to accept the waste for management.

I.3 Implementation Experience to Date

NRC ensures that exports and non-DOE imports of nuclear materials facilities and equipment under the Agency's jurisdiction are licensed in accordance with applicable U.S. statutory and regulatory requirements, as well as U.S. Government commitments towards legally binding international treaties and multilateral and bilateral agreements. In addition, NRC and DOE continue to exercise global leadership by adhering to and promoting the adoption of international guidance such as the Code of Conduct on the Safety and Security of Radioactive Sources. Examples of specific accomplishments include:

- Refining criteria for approving exports of Category 1 and 2 radioactive sources;
- Participating in international regulatory meetings to strengthen controls on transfers of radioactive sources of concern without disrupting legitimate commerce;
- Maintaining close relationships with our hemispheric neighbors and primary trading partners to continue refining export and import protocols, procedures and controls; and
- Participating in meetings of the Nuclear Suppliers Group.[201]

---

[201] The Nuclear Suppliers Group is a group of nuclear supplier countries, which seeks to contribute to the non-proliferation of nuclear weapons through the adherence to Guidelines for nuclear exports and nuclear related exports. See http://www.nuclearsuppliersgroup.org/.

130

U.S. Fourth National Report-Joint Convention on the Safety of Spent Fuel Management and on the Safety of Radioactive Waste Management

Table I-1 provides the total number of licensing actions and Table I-2 provides the number of specific licenses issued for import or export of Category 1 and 2 radioactive materials. Appendix P to 10 CFR Part 110 provides import and export threshold limits for specific radionuclides. This includes discrete sources of P226PRa. Note that the 2010 Part 110 amendment eliminated the specific license requirement for imports of Category 1 and 2 sources unless they are in the form of radioactive waste not excluded from the revised definition in 10 CFR 110.2.

| Table I-1 Completed Export/Import NRC Licensing Actions for January 2000 – July 2011 | | | | | | | | | | | | | |
Year	2000	2001	2002	2003	2004	2005	2006	2007	2008	2009	2010	2011	Total
Byproducts	1	0	0	2	6	7	2	3	1	1	7	3	33
Components	13	19	5	7	15	10	11	16	5	9	13	12	135
Moderator Material	4	2	3	3	6	3	1	2	1	2	1	2	30
Reactors & Major Reactor Components	4	5	2	0	1	3	1	4	1	3	0	0	24
Special Nuclear Material	50	86	68	53	56	51	64	52	43	69	41	29	662
Source Material	10	17	2	14	11	4	5	8	12	5	5	5	98
Waste Exports	2	3	1	4	1	5	2	5	3	0	6	3	35
Waste Imports	3	4	0	2	2	3	3	9	2	1	5	3	37
Total	87	136	81	85	98	86	89	99	68	90	78	57	1054

Source: NRC, Office of International Programs.

| Table I-2 Appendix P Licenses Issued December 2005 – July 2011 | | | | |
Year	Combination License	Export License	Import License	Total
2005	15	0	0	15
2006	49	15	19	83
2007	54	12	7	73
2008	21	12	6	39
2009	26	13	5	44
2010	26	12	5	43
2011	0	10	0	10
Total	191	74	42	307

Source: NRC, Office of International Programs

J. DISUSED SEALED SOURCES

J.1   General Safety for Sealed Sources

Radiation safety programs for use of byproduct material as a sealed source or device are based on robust containment of radioactive material. Sealed sources or devices are designed to withstand stresses imposed by the environment in which they are possessed and used. Regulations in 10 CFR Parts 30, 31, 32, 34, 35, 36, and 39 provide requirements for both vendors and users of sealed sources and devices. Agreement States issue adequate and compatible regulations for the control of sealed sources and devices within their borders.

Certain regulations require radioactive sources in products used under a specific license issued in accordance with 10 CFR Parts 30-39 to be registered with NRC or an Agreement State. In addition, for products using byproduct material in the form of a sealed source or in a device that contains the sealed source, 10 CFR 30.32(g) requires a license applicant to either make reference to a registered sealed source or device or provide the information necessary to perform a safety evaluation of the sealed source or device. 10 CFR 32.210 outlines NRC's safety evaluation and registration criteria; furthermore, the regulations clarify the regulatory responsibility of those who hold product registration certificates. This practice has been used since the 1950s and allows regulatory agencies to ensure designs meet all regulatory requirements, maintain their integrity under both normal use and credible accident conditions. This process allows applicants and license reviewers to reference the evaluation when licensing the product for use or distribution without having to perform a complete evaluation of the product for each licensing action. Regulations in 10 CFR Parts 34, 35, 36, and 39 provide additional requirements for specific types of sources and devices. Regulations in 10 CFR Parts 30, 31, and 32 also allow for use of equipment requiring registration but not requiring a license for use, and for sources and devices requiring neither registration nor licensing if they meet certain requirements.

NRC and the Agreement States perform safety evaluations of the ability of sealed sources and devices to contain licensed material for use under the conditions requested. These evaluations are summarized in registrations. The registrations are maintained by NRC in the National Sealed Source and Device Registry. Access to the Registry is currently unavailable for general public access; however, reports on current and former vendors and products can be downloaded by the general public. Information on the regulatory process for evaluating and licensing sealed sources and devices is available to the public through NRC's website.[202] Agreement States also provide information on their radiation safety evaluations to NRC for the registry. A vendor only needs to provide detailed information about its sealed source or device to the regulatory agency that has jurisdiction. NRC estimates there are approximately 2 million of these devices and sources in existence.

As part of the International Atomic Energy Agency (IAEA) implementation of the "Action Plan for the Safety of Radiation Sources and Security of Radioactive Material," in 2005, the IAEA developed the International Catalogue of Sealed Sources and Devices. The key objectives of this activity are to provide information to Member States to assist in the identification of disused or uncontrolled radioactive sources and devices government

---

[202] See http://www.nrc.gov/materials/miau/sealed-source.html.

133

U.S. Fourth National Report-Joint Convention on the Safety of Spent Fuel Management and on the Safety of Radioactive Waste Management

agencies and private entities may encounter and to provide readily available hazard information to assist and inform emergency response personnel. The Catalogue was initially populated with information provided by NRC from its National Sealed Source and Device Registry. The IAEA is continuing to populate the Catalogue from sealed source and device information from other Member States and periodic updates from NRC and Department of Energy (DOE). In addition, the IAEA Office of Nuclear Security, drawing on resources provided by donor countries to the Nuclear Security Fund, supports the security of in use and disused sources.

Licensees possessing, using, packaging, handling, transferring, and disposing licensed material are required to comply with the general occupational and public radiological protection regulations, listed in Table E-2. This includes licensing, financial assurance and record keeping for decommissioning, and expiration and termination of licenses and decommissioning.

J.2 Disused Sealed Sources Reentry from Abroad

NRC published regulations on November 8, 2006 (Federal Register, 71 FR 65685) to implement the National Source Tracking System (NSTS). The purpose is to enhance controls for certain radioactive materials considered to be of the greatest concern from a safety and security standpoint. The NSTS involves other Federal and State agencies and international partners and requires NRC licensees, and Agreement States to report all Category 1 and 2 sealed sources. The NSTS requires licensees to report the manufacture, transfer, receipt, disassembly, and disposal of nationally tracked sources. DOE also contributes information about DOE sealed sources to the NSTS. The NSTS is an important component of NRC's effort to enhance the control of radioactive material and prevent its use by the nation's adversaries. The NSTS enhances the ability of NRC and Agreement States to conduct inspections and investigations, communicate information to other government agencies, and verify legitimate ownership and use of nationally tracked sources. There are a little over 75000 of these sources in use. Section K.3 addresses plans to enhance the NSTS by linking Web-based licensing and the License Verification System (LVS) to constitute a comprehensive program for security and control of radioactive material.

U.S. regulations do not prohibit the return of disused sealed sources (see Section I.2.2). The U.S., recognizing the need to address the threat of radiological terrorism, has led international efforts to strengthen controls over international transfers of radioactive sources and materials, including those sources that could be used in a radioactive dispersal device or "dirty bomb." U.S. efforts have yielded significant progress, including the revision of the non-legally binding IAEA Code of Conduct on the Safety and Security of Radioactive Sources (Code). G-8 Leaders[203] agreed at the Sea Island Summit in June 2004 to work toward effective import/export controls for radioactive sources. The IAEA Board of Governors approved the IAEA Code of Conduct and the Import/Export Guidance [204] for Radioactive Sources by an IAEA expert group representing 41 Member

---

[203] The Group of Eight (G8), also known as Group of Seven and Russia, is an international forum for the governments of Canada, France, Germany, Italy, Japan, Russia, the United Kingdom and the U.S. The G8 can refer to the member states or to the annual summit meeting of the G8 heads of government.
[204] This guidance was endorsed by the IAEA General Conference on September 24, 2004 (see GC(48)/RES/10), and published by the IAEA on March 30, 2005 (IAEA/CODEOC/IMP-EXP/2005). Certain imports and exports for defense purposes are excluded from the IAEA Code of Conduct and the Import/Export guidance.

States. As of June 2011, 103 countries had made political commitment to the Code and 64 of these had also adopted the Guidance.

NRC amended domestic licensing requirements to be consistent with the revised Code and international import/export guidance for imports and exports of IAEA Category 1 and 2 radioactive sources and material. Beginning on January 1, 2006, transfers of these radioactive sources out of the U.S. must be approved by NRC in a specific export license as noted in Section I-2. In light of recent enhancements made to NRC's domestic regulatory framework, as of August 2010, a specific license is no longer needed for the import of this material; the activity can be performed under a general license although prior shipment notifications are still required. DOE is also developing guidance for sealed sources within its authority and control, consistent with the IAEA Code of Conduct and the Import/Export guidance. The U.S. continues to promote greater harmonization for strengthening controls over international transfers of radioactive sources.[205]

J.3   Sealed Sources Security

In the U.S., radioactive sealed sources, which make up less than 1 percent of all low-level radioactive waste by volume and activity, pose a national security concern because most licensees lack access to commercial disposal facilities for the sources when the sources become disused. The U.S. is actively engaged in addressing this security concern through the efforts described below.

The EPAct05 required establishment of the Interagency Task Force on Radiation Source Protection and Security (Task Force), which has made important progress since its original 2006 report to the President and Congress to improve the security of domestic radioactive sources.[206] The Task Force Report found that by far the most significant challenge is access to disposal for disused radioactive sources. Although this is a security initiative, the focus on obtaining a disposal path for these sources aligns with the safety goal of isolating these sources from the public and environment. The Task Force Report for 2010 provides an update to the status of radiation sources, and addresses possible disposal paths, alternate technologies and progress in control and accountability of disused sources. This report documents the success of programs such as the NSTS, DOE/National Nuclear Security Administration (NNSA) Global Threat Reduction Initiative recovery of disused sources and voluntary security upgrades for in-use sources.[207] Although the conclusions and recommendations in the 2010Task Force Report focus on system and infrastructure the Report also acknowledges ongoing activities associated with radiation detection efforts in transboundary shipments and traffic.

Retrieved sources are managed in accordance with the objectives of the Joint Convention found in Article 1. Disused sources are not waste until they are accepted for disposal at commercial or governmental facilities.

---

[205] Key objectives of the G-8 Evian, Sea Island and Glen Eagle Summits.

[206] Task Force Report (2006) is available at:
http://hps.org/govtrelations/documents/nrc_source_taskforce_report.pdf.

[207] The 2010 update of the Task Force Report is available at: http://www.nrc.gov/security/byproduct/2010-task-force-report.pdf.

135

Licensees possessing disused sealed sources, that are Greater-than-Class C LLW, are responsible for properly storing the sources in accordance with NRC or Agreement State regulatory and/or license requirements until a disposal facility or alternative disposition path (such as recycling) is available. These sources must be stored at the owners' facilities or at commercial facilities in accordance with NRC or Agreement State licensing requirements (10 CFR Parts 30, 32, 33, 34, 35, 36, and 39). These regulations require licensees to secure the material in accordance with regulatory requirements to prevent unauthorized removal or access licensed materials stored in controlled or unrestricted areas.

The U.S. has continued and expanded efforts to recover disused radioactive sealed sources. DOE/NNSA Global Threat Reduction Initiative mission is to reduce and protect high-risk nuclear and radiological materials located at civilian sites worldwide. DOE/NNSA recovers disused sealed sources in the possession of NRC and Agreement States licensees that present threats to national security, public health or safety. DOE/NNSA and NRC also support disused source recovery efforts implemented by the Conference of Radiation Control Program Directors (CRCPD). All sources recovered by these programs had been voluntarily registered by their owners as disused through DOE/NNSA.[208] Since October 2008, these programs combined have recovered nearly 10000 sources. As previously mentioned, DOE disposes of LLW, including sealed sources that have been determined to be waste, in accordance with DOE policy and requirements in DOE Order 435.1 and DOE Manual 435.1-1. In addition, the EPA has the authority and capability to recover orphan sources.

In September 2008, the U.S. Department of Homeland Security (DHS) facilitated a public and private sector Sealed Source Security Workshop. Workshop participants, including Federal, State, local and private sector stakeholders, identified the lack of a commercial disposal pathway for disused radioactive sealed sources as a national security concern. Following this meeting, DHS established a working group charged with characterizing the sealed source disposition challenge, developing a consensus problem statement, investigating and recommending medium- and long-term options and developing a messaging strategy for communicating with the appropriate stakeholders to implement a solution.

J.4 Illicit Trafficking and Nuclear Terrorism

The U.S. actively supports a wide range of international, multilateral, and bilateral initiatives to minimize the potential for unauthorized acquisition, possession, use, transfer and/or disposal of radioactive materials of concern to the global community. The U.S. contributes to programs, such as the Global Initiative to Combat Nuclear Terrorism, and urges all nations to adhere to legally binding commitments and follow non-binding guidance to enhance the safe and secure use of radioactive materials including sealed sources and devices covered by the IAEA Code of Conduct on the Safety and Security of Radioactive Sources.

The U.S. continues to work with numerous partners internationally, multilaterally and bilaterally to further develop and/or strengthen national as well as regional detection and response capabilities to deter illicit trafficking and unauthorized transfers. The U.S. also supports a wide range of collaborative efforts to increase the effectiveness of information sharing networks, including among border protection officials, so that responsible

---

[208] See http://osrp.lanl.gov/.

authorities are better able to recognize the circumstances, patterns and trends associated with such activities. In this regard, the IAEA's Illicit Trafficking Database, which was set up in 1995 to collect information on illicit trafficking and other unauthorized activities involving nuclear and radioactive material, is one of several tools that augments and supports the safety and security objectives of the IAEA Code of Conduct.

NRC has worked with DHS Customs and Border Protection, and a program is in place to verify the legitimacy of shipments of licensed radioactive material entering the U.S. through established ports of entry. Specific information about NRC's cooperation with U.S. Customs can be accessed from NRC's webpage on NRC's Source Data Team and U.S. Customs.[209]

The responsibility for developing a global detection system for illicit sources has been designated to the Domestic Nuclear Detection Office (DNDO) within DHS, which was established to develop, acquire, and support the deployment of a domestic system to detect and report attempts to import or transport a nuclear device or fissile or radiological material intended for illicit use. These efforts will increase the ability of law enforcement officials to detect licensed material in shipments and provide a means to intercept lost or stolen domestic sources or illegal shipments along major highway routes. The domestic system being deployed consists of handheld and portal radiation monitors purchased by state and local entities (typically, law enforcement and fire brigades). DNDO is developing and will provide training for the state and local entities.

---

[209] See http://www.nrc.gov/security/byproduct/export-import/source-data-team.html.

K. PLANNED ACTIVITIES TO IMPROVE SAFETY

This report describes ongoing U.S. activities ensuring safe management of spent fuel, radioactive waste, and disused sealed sources. There are several key areas important to safety continuing to receive much attention.

K.1   Spent Fuel and HLW Disposition

The Blue Ribbon Commission on America's Nuclear Future will provide recommendations for developing a safe, long-term solution to managing the Nation's used nuclear fuel and nuclear waste. Following the release of the final report in January 2012, the U.S. will proceed with developing a new policy for managing spent fuel and HLW.

DOE continues to support research and technology development for long-term solutions and for storage, transportation, and disposal of spent fuel and wastes generated by existing and future nuclear fuel cycles. In the meantime, commercial reactor sites will continue to store spent fuel in reactor pools and dry casks using NRC approved designs.

K.2   Commercial LLW Disposal

Some uncertainty exists about the availability of commercial disposal for Class B and C LLW. The planned commissioning of the new Texas Compact disposal facility may eliminate this uncertainty for LLW generators that currently do not have a disposal path for Class B and Class C LLW. As discussed in Section A.5.2, industry has devised other innovative approaches to mitigate disposal challenges. The U.S. continues to monitor commercial LLW disposal capacity and projected needs.

K.3   Disused Sealed Sources and Greater-Than-Class C LLW Disposal

The U.S. is engaged in various efforts to ensure the safety and security of disused sealed sources. NRC, for example, implemented regulatory changes strengthening domestic licensing requirements for import and export of high-risk radioactive sources and materials. These revisions to 10 Code of Federal Regulations (CFR) Part 110 bring U.S. import/export controls in line with the revised IAEA Code of Conduct on the Safety and Security of Radioactive Sources and international import/export guidance. More recently, Part 110 was amended to clarify that U.S.-origin sealed sources are excluded from the definition of radioactive waste. There is an ongoing effort to find a disposal path for these sources.

The NRC is in the process of establishing the Integrated Source Management System[210]. This system enhances the National Source Tracking System (NSTS)[211] by linking Web-based Licensing (WBL), and the License Verifications System (LVS). These constitute a comprehensive program for the security and control of radioactive material. When complete, these components will form an integrated source management system that will include information on all NRC and Agreement State licensees and over 50000 high-risk radioactive sources possessed by approximately 1,300 licensees. The WBL is

---

[210] Vision for Integrated Source Management http://www.nrc.gov/security/byproduct/nsts/vision-source-manage.html.
[211] Overview of the NSTS http://www.nrc.gov/security/byproduct/nsts.html.

139

U.S. Fourth National Report-Joint Convention on the Safety of Spent Fuel Management and on the Safety of Radioactive Waste Management

expected to be deployed by late 2012, the LVS is expected to be deployed by early 2013.

The benefits of system integration include:
- Make national radioactive source authorization, possession and transaction information available to other government agencies with a role in the protection of the Nation from nuclear and radiological threats.
- Provide licensees with a secure automated means to verify license information and possession authorization prior to initiating radioactive material transfers.
- Enable NRC to monitor the location, possession, transfer and disposal of high-risk radioactive sources through the country.
- Improve source accountability, and alert the regulators to tracking discrepancies.
- Modernize NRC licensing and inspection management systems.

As discussed in Section D.2.1.1, DOE is analyzing alternatives for the disposal of GTCC LLW and DOE GTCC-like waste. DOE plans to issue a final EIS in 2012. As required by EPAct05, DOE will submit a report to the U.S. Congress and await Congressional action before reaching a final decision on a disposal option(s) for GTCC LLW.

K.4 DOE Waste Treatment Facilities

Large multi-year construction projects are ongoing at the DOE Hanford Site, Idaho Site, and Savannah River Site (SRS) to ultimately retrieve and treat tank waste so that underground tanks can be decommissioned and closed, and tank waste can be safely dispositioned. These facilities are key to cleanup of legacies from historical defense activities in the U.S. DOE/NNSA is constructing a Waste Solidification Building (WSB) at SRS, to process certain future waste streams from the Mixed Oxide Fuel Fabrication Facility and pit disassembly and conversion operations into a solid form for ultimate disposal. The WSB must be operational to support mixed oxide (MOX) cold start-up testing activities. See Section D.2.1.3 for additional information.

K.5 Regulation of Residual Contamination at Military Sites

NRC has moved to the regulation of military sites contaminated with depleted uranium from past testing of munitions. The EPAct05 has resulted in NRC's regulatory authority over discrete sources of ^{226}Ra and accelerator-produced radioactive material, including some military ^{226}Ra. There is still a jurisdictional issue over what military ^{226}Ra should fall under NRC's regulatory authority. This effort is still in its early stages, which includes completing an inventory of such military sites in the U.S., as well as determining the extent of contamination. This involves the various branches of the U.S. military services. Other radionuclides that may have been disposed of include ^{90}Sr.

K.6 NRC Rulemaking Activities

NRC maintains a regulatory agenda of petitions for rulemaking activities and other planned rulemaking.[212] It contains a summary and the status for each ongoing rulemaking and petition for rulemaking received by the agency. Some rulemaking activities relating to the provisions in the Joint Convention are:

---

[212] This information is updated semiannually at http://www.nrc.gov/reading-rm/doc-collections/nuregs/staff/sr0936/.

140

U.S. Fourth National Report-Joint Convention on the Safety of Spent Fuel Management and on the Safety of Radioactive Waste Management

1. Decommissioning Planning — This rule amends NRC regulations to prevent future legacy sites, i.e., facilities in decommissioning with complex issues and owners incapable of completing decommissioning because of financial or technical difficulties. One set of changes revises 10 CFR 20.1406 and 20.1501 to require that licensees conduct their operations to minimize the introduction of residual radioactivity at the site, including the subsurface, and to keep contamination survey results with records important for decommissioning. A second set of changes revises regulations in 10 CFR Parts 30, 40, 50, 70, and 72 to provide tighter control of decommissioning financial assurances and more detailed reporting by licensees of their decommissioning cost estimates. The final rule was published in the Federal Register at 76 FR 35512 on June 17, 2011 and will take effect in January 2013, except for certain financial assurance status reporting provisions, which take effect in March 2013.

2. Consideration of Environmental Impacts of Temporary Storage of Spent Fuel After Cessation of Reactor Operation and Waste Confidence Decision Update (Part 51) — The final rule amended the Commission's regulations by revising its generic determinations on the timing of the availability of a geologic repository for commercial HLW and spent fuel and on the environmental impacts of storage of spent fuel at or away from reactor sites after the expiration of reactor operating licensing. The proposed rule was published October 2008; the public comment period closed December 2008, but was extended until February 2009. The final rule was published in the Federal Register at 75 FR 81032 in December 2010 and was effective on January 24, 2011.

3. Advance Notification to Native American Tribes of Transportation of Certain Types of Nuclear Waste — The rule proposes to amend NRC regulations that govern packaging and transportation of radioactive material and physical protection of plants and materials. Specifically, the proposed amendments will require licensees to provide advance notification to Federally recognized Tribal governments about shipments of irradiated reactor fuel and certain nuclear wastes for any shipment that passes within or across their reservations. The Tribal government would be required to protect the shipment information as Safeguards Information. The proposed rule was published in the Federal Register at 75 FR 75641 on December 6, 2010; the public comment period closed February 2011. The final rule is still in process.

4. Part 72 License and Certificate of Compliance Terms — The final rule will amend the regulations for licensing requirements for the independent storage of spent fuel, HLW, and reactor-related GTCC LLW. The amendments will clarify the terms for approved dry spent fuel storage cask designs, also known as Certificates of Compliance (CoCs), and independent spent fuel storage installation (ISFSI) licenses. Specifically, this final rule extends the initial and renewal license terms for site-specific ISFSI licenses and CoCs from a term not to exceed 20 years to a term not to exceed 40 years. License and CoC renewal applications will need to include an analysis that considers the effects of aging on structures, systems, and components important to safety for the requested renewal term. The final rule allows general licensees to implement changes authorized by an amended CoC to a cask loaded under the initial CoC or an earlier amended CoC (a "previously loaded cask"). The final rule was published

141

U.S. Fourth National Report-Joint Convention on the Safety of Spent Fuel Management and on the Safety of Radioactive Waste Management

in the Federal Register at 76 FR 8872 on February 16, 2011. Implementation will follow.

5. Site-Specific Analyses Rule Making for Near Surface Disposal (10 CFR Part 61)
   NRC is considering new regulations on near surface disposal of waste streams not contemplated in the development of 10 CFR Part 61, including significant quantities of depleted uranium. NRC is considering amending 10 CFR Part 61 to require operating and future LLW disposal facilities to conduct site-specific analyses to demonstrate compliance with performance objectives in 10 CFR Part 61. These amendments would enhance the safe disposal of low-level radioactive waste and would also identify any additional measures that would be prudent to implement. The changes would address site-specific analyses for the disposal of waste streams not contemplated during the development of 10 CFR Part 61, such as large quantities of depleted uranium, blended waste and reprocessing waste, and the technical requirements for such an analysis. A guidance document will be prepared for public comment outlining approaches that can be used in the site-specific analyses. NRC is also considering additional changes to the current regulations to reduce ambiguity, facilitate implementation, and to better align the requirements with current health and safety standards. This regulation would apply to existing and future LLW disposal facilities that are regulated by NRC and the Agreement States. The proposed rule was published for public review and comment in the Federal Register at 76 FR 24831 on May 3, 2011.

This rulemaking is part of an overall strategy to consider a potential revision of the 10 CFR Part 61 regulations. Public workshops were held to gather information from a broad spectrum of stakeholders concerning NRC's proposed options for a comprehensive revision of 10 CFR Part 61.[213]

K.7 International Collaboration

Establishing and maintaining an international program is an important part of the U.S. vision of safe and secure disposal of spent fuel, because working with other countries expands the technical depth of individual programs, offering knowledge in optimizing disposal systems. International activities link the U.S. to the world's evolving spent fuel and high-level waste management practices, and provide a forum for exchanging strategies and technologies with other nations. Through international cooperation, countries can coordinate on the development of formal agreements with each other, enabling an exchange of detailed scientific and technical information, or joint sponsorship of activities in areas such as alternatives for geologic disposal environments and long-term storage.

Participation in these types of activities benefits the U.S. through the acquisition and exchange of information, particularly related to complex issues dealing with topics such as performance assessment, database development, and peer review by experts of other participating nations. These international projects serve U.S. goals in advancing scientific understanding, enhancing environmental protection, and improving global safety and security. In fostering international cooperation on spent fuel and radioactive waste management and disposal, the goal is to lead to an optimized national disposal

---

[213] More specific information can be accessed at the NRC's website at: http://www.nrc.gov/waste/llw-disposal/public-outreach.html.

142

U.S. Fourth National Report-Joint Convention on the Safety of Spent Fuel Management and on the Safety of Radioactive Waste Management

system and promote the exchange of institutional and technical knowledge throughout the international community.

There is general consensus that deep geologic disposal is the preferred option for permanent management of spent fuel and HLW. Consequently, international collaboration with other countries is an important aspect that would allow for increased interactions and sharing of knowledge including: years of world-class science in site investigations and characterization; development of integrated radioactive waste management systems; experience in transporting nuclear materials; opportunities to enhance science on nuclear waste disposition; and innovative technologies and methods.

The U.S. works with other countries internationally on radioactive waste management programs. This collaboration is handled through several venues and organizations. The following list includes a short description of some of the groups with whom the U.S. has continued interactions:

- Organization for Economic Cooperation and Development/Nuclear Energy Agency (OECD/NEA), Radioactive Waste Management Committee (RWMC). The U.S. delegation participates in annual meetings, provides status reports on the country's radioactive waste and decommissioning programs, and reviews/updates international reports prepared as reference documents.

- International Association for Environmentally Safe Disposal of Radioactive Materials (EDRAM). EDRAM is an association of executives and chairmen of worldwide radioactive waste management organizations that meets biannually to share and compare program information among the eleven member countries: Belgium, Canada, Finland, France, Germany, Japan, Spain, Sweden, Switzerland, United Kingdom, and United States of America.

- The Extended Storage Collaboration Program (ESCP). The ESCP is a consortium of organizations investigating aging effects and mitigation options for the extended storage of spent fuel, followed by transportation. Over the next year, potential research, and identification of available suitable demonstration facilities in these countries will be pursued. The consortium is also working to identify all relevant information for extended storage that has already been produced worldwide. See Section A.5.1.

- International Framework for Nuclear Energy Cooperation (IFNEC),[214] formerly the international portion of the Global Nuclear Energy Partnership. IFNEC is a forum through which participating states explore mutually beneficial approaches to ensure the use of nuclear energy for peaceful purposes proceeds in a manner that is efficient and meets the highest standards of safety, security and non-proliferation. Participating states do not give up any rights and voluntarily engage in information sharing:

  o Reliable Nuclear Fuel Services Working Group — seeks to identify common interests among the participant countries and recommend practical measures for moving towards reliable comprehensive fuel service arrangements, including front-end services and spent fuel management.

---

[214] For more information on IFNEC, see www.ifnec.org.

- o Infrastructure Development Working Group — supports the safe and secure development of nuclear energy infrastructure, including infrastructure needs for international nuclear fuel service frameworks. It has a Sub-Group on Radioactive Waste Management, which aims to share information, experiences, and broader knowledge that could inform countries' development of radioactive waste management positions and strategies.

- International Atomic Energy Agency (IAEA). The IAEA sponsors technical meetings, some of which pertain to different topics related to radioactive waste management such as long-term storage, transportation, safety, etc. Technical experts participate in, and/or monitor the progress of these meetings, depending on the topic and expertise needed.

- Multi-lateral initiatives. The U.S. supports and participates in international cooperative interactions, such as the Joint Standing Committee on Nuclear Energy Cooperation (JSCNEC) with Argentina, Brazil and Korea.

- Bilateral agreements and Memoranda of Understanding. The U.S. uses these mechanisms with other countries to establish and address technical cooperative activities for radioactive waste management activities of mutual interest. Cooperation between the parties is based on mutual benefit, equality, and reciprocity. The agreements could involve cooperation which might include, but not limited to: exchange of scientists and engineers; exchange of scientific and technical information and results of R&D; exchange of samples and materials for testing; organization of and participation in seminars and meetings on agreed upon topics; visits to specific facilities of interest; and joint projects for topics of mutual agreement.

K.8 Waste Disposition for Commercial Medical Isotope Production

The DOE/National Nuclear Security Administration (NNSA) is working to accelerate commercial production of the medical isotope molybdenum-99 (Mo-99) in the U.S., without the use of highly enriched uranium (HEU). DOE/NNSA is supporting the U.S. private sector to accelerate independent, non-HEU-based technical pathways to produce Mo-99 in the U.S. in cooperation with commercial partners and the U.S. national laboratories. The NRC or Agreement State would have to license any new commercial production facility. The objective of each of these projects is to produce non-HEU-based Mo-99 by the end of 2014. The expected waste streams from the production of Mo-99 are likely to include radioactive waste for which there is currently no commercial disposal path. The projects are currently under development, and production has not yet commenced at the time this report was written. However, disposition of waste and spent nuclear fuel considerations impact the technical and economic viability of each of the projects. Until a disposal path is identified, producers of this medical isotope would need to provide on-site storage.

ANNEXES

145

U.S. Fourth National Report-Joint Convention on the Safety of Spent Fuel Management and on the Safety of Radioactive Waste Management

Annex D-1A Spent Fuel Management Facilities[215] Government Facilities

State	Installation	Facility	Function	Licensee	Regulator	SF[216] Source	Inventory	Units	Estimated Activity (Bq)
Colorado	U.S. Geological Survey (Denver)	Research/Test Reactor	Wet Storage	U.S. Geological Survey	NRC	2	0.04	MTHM	5.37E+14
	Fort St. Vrain	ISFSI	Dry Storage	DOE	NRC	2	14.73	MTHM	1.73E+17
Idaho	Idaho National Lab	INTEC-666	Wet Storage	DOE	DOE	1, 2	2.93	MTHM	5.96E+17
		Multiple INL facilities	Dry Storage	DOE	DOE	1,2	195.74	MTHM	1.22E+17
		ISFSI[217]	Dry Storage	DOE	NRC	2	81.59	MTHM	1.84E+18
Illinois	Argonne National Lab	ANL SF Storage	Dry Storage	DOE	DOE	1, 2	0.12	MTHM	4.11E+15
Maryland	National Institute of Standards and Technology (Gaithersburg)	Research/Test Reactor	Wet Storage	National Institute of Standards and Technology	NRC	2	0.02	MTHM	7.88E+06
	Armed Forces Radiobiology Research Institute (Bethesda)	Research/Test Reactor	Wet Storage	Armed Forces Radiobiology Research Institute	NRC	1	0.02	MTHM	3.02E+14
New Mexico	Sandia National Lab	Multiple SNL Facilities	Dry Storage	DOE	DOE	1, 2	0.24	MTHM	7.40E+16
Rhode Island	Rhode Island Atomic Energy Commission (Narragansett)	Research/Test Reactor	Wet Storage	Rhode Island Atomic Energy Commission	NRC	2	0.02	MTHM	3.96E+14
South Carolina	Savannah River Site	L-Basin	Wet Storage	DOE	DOE	1, 2	29.42	MTHM	1.74E+18
			Dry Storage	DOE	DOE	2	0.01	MTHM	3.40E+15
Tennessee	Oak Ridge Reservation	Multiple ORR Facilities	Wet Storage	DOE	DOE	2	0.47	MTHM	6.73E+17
Washington	Hanford Site	Multiple Hanford facilities	Dry Storage	DOE	DOE	1, 2	2127.88	MTHM	2.01E+18
		K Basin[218]	Wet Storage	DOE	DOE	1	1.50	MTHM	9.95E+14

---

215 Data Source: DOE National Spent Fuel Program database as of 12/31/2010.

216 SF Sources: 1-Defense applications; 2-Commercial NPPs and Test/Research Reactors.

217 In addition to this facility, NRC licensed a second DOE ISFSI at Idaho, which DOE subsequently decided not to construct.

218 K Basin material includes Knock Out Pot material, fuel found in burial grounds, and fuel fragments in the pool

U.S. Fourth National Report-Joint Convention on the Safety of Spent Fuel Management and on the Safety of Radioactive Waste Management

Annex D-1B Spent Fuel Management Facilities University Research Facilities[219]

State	Installation	Facility	Function	Licensee	Regulator	Inventory	Units	Estimated Activity (Bq)
California	University of California (Irvine)	Research Reactor	Wet Storage	University of California	NRC	21.42	kgU	9.99E+13
	University of California (Davis)[220]	Research Reactor	Wet Storage	University of California	NRC	72.31	kgU	9.95E+14
Florida	University of Florida (Gainesville)	Research Reactor	Wet Storage	University of Florida	NRC	23.81	kgU	1.38E+14
Indiana	Purdue University (West Lafayette)	Research Reactor	Wet Storage	Purdue University	NRC	17.53	kgU	`
Kansas	Kansas State University (Manhattan)	Research Reactor	Wet Storage	Kansas State University	NRC	21.08	kgU	7.70E+14
Maryland	University of Maryland (College Park)	Research Reactor	Wet Storage	University of Maryland	NRC	16.35	kgU	4.92E+14
Massachusetts	University of Lowell (Lowell)	Research Reactor	Wet Storage	University of Lowell	NRC	10.13	kgU	1.74E+14
	Massachusetts Institute of Technology (Cambridge)	Research Reactor	Wet Storage	Massachusetts Institute of Technology	NRC	19.85	kgU	3.36E+14
	Worcester Polytechnic Institute (Worcester)	Research Reactor	Wet Storage	Worcester Polytechnic Institute	NRC	22.75	kgU	1.41E+13
Missouri	University of Missouri (Columbia)	Research Reactor	Wet Storage	University of Missouri	NRC	35.44	kgU	5.03E+16
	University of Missouri (Rolla)	Research Reactor	Wet Storage	University of Missouri	NRC	26.46	kgU	3.29E+15
North Carolina	North Carolina State University (Raleigh)	Research Reactor	Wet Storage	North Carolina State University	NRC	315.40	kgU	6.22E+14
Ohio	Ohio State University (Columbus)	Research Reactor	Wet Storage	Ohio State University	NRC	26.15	kgU	3.16E+14
Oregon	Oregon State University (Corvallis)	Research Reactor	Wet Storage	Oregon State University	NRC	77.06	kgU	8.70E+14
	Reed College (Portland)	Research Reactor	Wet Storage	Reed College	NRC	30.86	kgU	4.37E+14
Pennsylvania	Pennsylvania State University (University Park)	Research Reactor	Wet Storage	Pennsylvania State University	NRC	37.57	kgU	1.88E+15
Texas	Texas A&M University (College Station)	Research Reactor (2)	Wet Storage	Texas A&M University	NRC	70.76	kgU	1.30E+15
	University of Texas (Austin)	Research Reactor	Wet Storage	University of Texas	NRC	35.17	kgU	2.68E+14
Utah	University of Utah (Salt Lake City)	Research Reactor	Wet Storage	University of Utah	NRC	26.82	kgU	1.22E+15
Washington	Washington State University (Pullman)	Research Reactor	Wet Storage	Washington State University	NRC	67.48	kgU	3.06E+15
Wisconsin	University of Wisconsin (Madison)	Research Reactor	Wet Storage	University of Wisconsin	NRC	68.06	kgU	4.37E+14

[219] Data Source: DOE National Spent Fuel Program database as of 12/31/2010.
[220] Formerly McClellan AFB (Sacramento)

Annex D-1C Spent Fuel Management Facilities Other Research and Nuclear Fuel Cycle Facilities
(Inventory as of December 31, 2010, per DOE projections)

State	Installation	Facility	Function	Licensee	Regulator	Inventory	Units	Estimated Activity (Bq)
California	Aerotest Research (San Ramon)	Research/Test Reactor	Wet Storage	Aerotest Research	NRC	17.43	kgU	3.54E+15
	General Electric (Pleasanton)	Research/Test Reactor	Wet Storage	General Electric	NRC	3.86	kgU	3.85E+12
Michigan	Dow Chemical Co. (Midland)	Research/Test Reactor	Wet Storage	Dow Chemical Co.	NRC	14.81	kgU	2.23E+14
Virginia	BWX Technology, Inc.	Fuel Cycle Facility	Dry storage	BWX Technology, Inc.	NRC	101.5	kgU	5.79E+15

U.S. Fourth National Report-Joint Convention on the Safety of Spent Fuel Management and on the Safety of Radioactive Waste Management

Annex D-1D Spent Fuel Management Facilities On Site Storage at Nuclear Power Plants (Inventory as of December 31, 2010, per DOE projections)[221]

State	Installation	Facility	Function	Licensee	Regulator	Inventory	Units
	Browns Ferry	ISFSI	Dry Storage	Tennessee Valley Authority	NRC	310	MTHM
Alabama	Browns Ferry 1, 2 & 3	Nuclear Power Plant	Wet Storage	Tennessee Valley Authority	NRC	1480	MTHM
	Farley	ISFSI	Dry Storage	Southern Nuclear Operating Co.	NRC	170	MTHM
	Farley 1 & 2	Nuclear Power Plant	Wet Storage	Southern Nuclear Operating Co.	NRC	1030	MTHM
Arizona	Palo Verde	ISFSI	Dry Storage	Arizona Public Service Co.	NRC	810	MTHM
	Palo Verde 1, 2 & 3	Nuclear Power Plant	Wet Storage	Arizona Public Service Co.	NRC	1080	MTHM
Arkansas	Arkansas Nuclear 1 & 2	Nuclear Power Plant	Wet Storage	Entergy	NRC	520	MTHM
	Arkansas Nuclear One	ISFSI	Dry Storage	Entergy	NRC	720	MTHM
	Diablo Canyon 1 & 2	ISFSI	Dry Storage	Pacific Gas & Electric Co.	NRC	220	MTHM
	Diablo Canyon 1 & 2	Nuclear Power Plant	Wet Storage	Pacific Gas & Electric Co.	NRC	880	MTHM
	Humboldt Bay	ISFSI	Dry Storage	Pacific Gas & Electric Co.	NRC	30	MTHM
California	Humboldt Bay (shutdown)	Nuclear Power Plant	Wet Storage	Pacific Gas & Electric Co.	NRC	30	MTHM
	Rancho Seco (shutdown)	ISFSI	Dry Storage	Sacramento Municipal Utility District	NRC	230	MTHM
	San Onofre	ISFSI	Dry Storage	Southern California Edison Co.	NRC	370	MTHM
	San Onofre 1 (shutdown), San Onofre 2 & 3	Nuclear Power Plant	Wet Storage	Southern California Edison Co.	NRC	1090	MTHM
	Haddam Neck (shutdown)	ISFSI	Dry Storage	Connecticut Yankee Atomic Power Co.	NRC	410	MTHM
Connecticut	Millstone	ISFSI	Dry Storage	Dominion Generation	NRC	180	MTHM
	Millstone 1 (shutdown), Millstone 2 & 3	Nuclear Power Plant	Wet Storage	Dominion Generation	NRC	1350	MTHM

[221] These numbers are estimates. The most recently collected RW-859 Nuclear Fuel Data survey included data to December 31, 2002. Updates since then are based on DOE projections and publicly available information.

150

U.S. Fourth National Report-Joint Convention on the Safety of Spent Fuel Management and on the Safety of Radioactive Waste Management

Annex D-1D Spent Fuel Management Facilities On Site Storage at Nuclear Power Plants (Inventory as of December 31, 2010, per DOE projections)[221]

State	Installation	Facility	Function	Licensee	Regulator	Inventory	Units
Florida	Crystal River 3	Nuclear Power Plant	Wet Storage	Progress Energy	NRC	540	MTHM
	St. Lucie	ISFSI	Dry Storage	FPL Group (including Florida Power & Light Co., NextEra Energy Resources)	NRC	140	MTHM
	St. Lucie 1 & 2	Nuclear Power Plant	Wet Storage	FPL Group (including Florida Power & Light Co., NextEra Energy Resources)	NRC	1070	MTHM
	Turkey Point 3 & 4	Nuclear Power Plant	Wet Storage	FPL Group (including Florida Power & Light Co., NextEra Energy Resources)	NRC	1130	MTHM
Georgia	Hatch	ISFSI	Dry Storage	Southern Nuclear Operating Co.	NRC	550	MTHM
	Hatch 1 & 2	Nuclear Power Plant	Wet Storage	Southern Nuclear Operating Co.	NRC	820	MTHM
	Vogtle 1 & 2	Nuclear Power Plant	Wet Storage	Southern Nuclear Operating Co.	NRC	1080	MTHM
Illinois	Braidwood 1 & 2	Nuclear Power Plant	Wet Storage	Exelon Generation Co.	NRC	1010	MTHM
	Byron	ISFSI	Dry Storage	Exelon Generation Co.	NRC	80	MTHM
	Byron 1 & 2	Nuclear Power Plant	Wet Storage	Exelon Generation Co.	NRC	1020	MTHM
	Clinton 1	Nuclear Power Plant	Wet Storage	Exelon Generation Co.	NRC	500	MTHM
	Dresden	ISFSI	Dry Storage	Exelon Generation Co.	NRC	560	MTHM
	Dresden 1 (shutdown), Dresden 2 & 3	Nuclear Power Plant	Wet Storage	Exelon Generation Co.	NRC	910	MTHM
	GE Morris[2]	ISFSI	Wet Storage	General Electric Co.	NRC	670	MTHM
	LaSalle County	ISFSI	Dry Storage	Exelon Generation Co.	NRC	40	MTHM
	LaSalle County 1 & 2	Nuclear Power Plant	Wet Storage	Exelon Generation Co.	NRC	1130	MTHM
	Quad Cities	ISFSI	Dry Storage	Exelon Generation Co.	NRC	300	MTHM
	Quad Cities 1 & 2	Nuclear Power Plant	Wet Storage	Exelon Generation Co.	NRC	1120	MTHM

Annex D-1D Spent Fuel Management Facilities On Site Storage at Nuclear Power Plants
(Inventory as of December 31, 2010, per DOE projections)[221]

State	Installation	Facility	Function	Licensee	Regulator	Inventory	Units
	Zion 1 & 2 (shutdown)	Nuclear Power Plant	Wet Storage	Exelon Generation Co.	NRC	1020	MTHM
Iowa	Duane Arnold	ISFSI	Dry Storage	FPL Group (including Florida Power & Light Co., NextEra Energy Resources)	NRC	110	MTHM
	Duane Arnold	Nuclear Power Plant	Wet Storage	FPL Group (including Florida Power & Light Co., NextEra Energy Resources)	NRC	370	MTHM
Kansas	Wolf Creek 1	Nuclear Power Plant	Wet Storage	Wolf Creek Nuclear Operating Corporation	NRC	610	MTHM
	River Bend	ISFSI	Dry Storage	Entergy	NRC	180	MTHM
Louisiana	River Bend 1	Nuclear Power Plant	Wet Storage	Entergy	NRC	390	MTHM
	Waterford 3	Nuclear Power Plant	Wet Storage	Entergy	NRC	600	MTHM
Maine	Maine Yankee (shutdown)	ISFSI	Dry Storage	Maine Yankee Atomic Power Co.	NRC	540	MTHM
Maryland	Calvert Cliffs	ISFSI	Dry Storage	Constellation Nuclear	NRC	690	MTHM
	Calvert Cliffs 1 & 2	Nuclear Power Plant	Wet Storage	Constellation Nuclear	NRC	540	MTHM
	Pilgrim 1	Nuclear Power Plant	Wet Storage	Entergy	NRC	520	MTHM
Massachusetts	Yankee Rowe (shutdown)	ISFSI	Dry Storage	Yankee Atomic Electric	NRC	130	MTHM
	Big Rock Point (shutdown)	ISFSI	Dry Storage	Entergy Nuclear Operations	NRC	60	MTHM
	D.C. Cook 1 & 2	Nuclear Power Plant	Wet Storage	Indiana Michigan Power Co.	NRC	1380	MTHM
Michigan	Enrico Fermi 2	Nuclear Power Plant	Wet Storage	Detroit Edison Co.	NRC	500	MTHM
	Palisades	ISFSI	Dry Storage	Entergy	NRC	400	MTHM
	Palisades	Nuclear Power Plant	Wet Storage	Entergy	NRC	200	MTHM

Annex D-1D Spent Fuel Management Facilities On Site Storage at Nuclear Power Plants (Inventory as of December 31, 2010, per DOE projections)[221]

State	Installation	Facility	Function	Licensee	Regulator	Inventory	Units
Minnesota	Monticello	ISFSI	Dry Storage	Northern States Power Co.-Minnesota	NRC	107	MTHM
	Monticello	Nuclear Power Plant	Wet Storage	Northern States Power Co.-Minnesota	NRC	220	MTHM
	Prairie Island	ISFSI	Dry Storage	Northern States Power Co.-Minnesota	NRC	420	MTHM
	Prairie Island 1 & 2	Nuclear Power Plant	Wet Storage	Northern States Power Co.-Minnesota	NRC	390	MTHM
Mississippi	Grand Gulf	ISFSI	Dry Storage	Entergy	NRC	140	MTHM
	Grand Gulf 1	Nuclear Power Plant	Wet Storage	Entergy	NRC	650	MTHM
Missouri	Callaway 1	Nuclear Power Plant	Wet Storage	Ameren	NRC	670	MTHM
Nebraska	Cooper Station	ISFSI	Dry Storage	Nebraska Public Power District	NRC	80	MTHM
	Cooper Station	Nuclear Power Plant	Wet Storage	Nebraska Public Power District	NRC	320	MTHM
	Fort Calhoun	ISFSI	Dry Storage	Omaha Public Power District	NRC	120	MTHM
	Fort Calhoun	Nuclear Power Plant	Wet Storage	Omaha Public Power District	NRC	270	MTHM
New Hampshire	Seabrook	ISFSI	Dry Storage	FPL Group (including Florida Power & Light Co., NextEra Energy Resources)	NRC	90	MTHM
	Seabrook 1	Nuclear Power Plant	Wet Storage	FPL Group (including Florida Power & Light Co., NextEra Energy Resources)	NRC	390	MTHM
New Jersey	Hope Creek/Salem 1&2	Nuclear Power Plant	Wet Storage	PSEG Nuclear LLC	NRC	1560	MTHM
	Hope Creek/Salem[222]	ISFSI	Dry Storage	PSEG Nuclear LLC	NRC	250	MTHM
	Oyster Creek	ISFSI	Dry Storage	Exelon Generation Co.	NRC	200	MTHM

222 Hope Creek and Salem nuclear power plants have a combined ISFSI.

Annex D-1D Spent Fuel Management Facilities On Site Storage at Nuclear Power Plants (Inventory as of December 31, 2010, per DOE projections)[221]

State	Installation	Facility	Function	Licensee	Regulator	Inventory	Units
	Oyster Creek	Nuclear Power Plant	Wet Storage	Exelon Generation Co.	NRC	410	MTHM
	Fitzpatrick	ISFSI	Dry Storage	Entergy	NRC	180	MTHM
	Fitzpatrick	Nuclear Power Plant	Wet Storage	Entergy	NRC	440	MTHM
	Ginna	ISFSI	Dry Storage	Constellation Nuclear	NRC	50	MTHM
New York	Ginna	Nuclear Power Plant	Wet Storage	Constellation Nuclear	NRC	380	MTHM
	Indian Point	ISFSI	Dry Storage	Entergy	NRC	190	MTHM
	Indian Point 1 (shutdown), Indian Point 2 & 3	Nuclear Power Plant	Wet Storage	Entergy	NRC	1030	MTHM
	Nine Mile Point 1 & 2	Nuclear Power Plant	Wet Storage	Constellation Nuclear	NRC	1080	MTHM
	Brunswick	ISFSI	Dry Storage	Progress Energy	NRC	40	MTHM
	Brunswick 1 & 2	Nuclear Power Plant	Wet Storage	Progress Energy	NRC	780	MTHM
North Carolina	McGuire	ISFSI	Dry Storage	Duke Power Co.	NRC	450	MTHM
	McGuire 1 & 2	Nuclear Power Plant	Wet Storage	Duke Power Co.	NRC	1017	MTHM
	Shearon Harris 1	Nuclear Power Plant	Wet Storage	Progress Energy	NRC	1140	MTHM
	Davis-Besse	ISFSI	Dry Storage	FirstEnergy Nuclear Operating Co.	NRC	30	MTHM
Ohio	Davis-Besse 1	Nuclear Power Plant	Wet Storage	FirstEnergy Nuclear Operating Co.	NRC	490	MTHM
	Perry 1	Nuclear Power Plant	Wet Storage	FirstEnergy Nuclear Operating Co.	NRC	590	MTHM
Oregon	Trojan (shutdown)	ISFSI	Dry Storage	Portland General Electric	NRC	360	MTHM
	Beaver Valley 1 & 2	Nuclear Power Plant	Wet Storage	FirstEnergy Nuclear Operating Co.	NRC	1010	MTHM
Pennsylvania	Limerick	ISFSI	Dry Storage	Exelon Generation Co.	NRC	130	MTHM
	Limerick 1 & 2	Nuclear Power Plant	Wet	Exelon Generation Co.	NRC	1070	MTHM

U.S. Fourth National Report-Joint Convention on the Safety of Spent Fuel Management and on the Safety of Radioactive Waste Management

Annex D-1D Spent Fuel Management Facilities On Site Storage at Nuclear Power Plants
(Inventory as of December 31, 2010, per DOE projections)[221]

State	Installation	Facility	Function	Licensee	Regulator	Inventory	Units
			Storage				
	Peach Bottom	ISFSI	Dry Storage	Exelon Generation Co.	NRC	600	MTHM
	Peach Bottom 2 & 3	Nuclear Power Plant	Wet Storage	Exelon Generation Co.	NRC	1050	MTHM
	Susquehanna	ISFSI	Dry Storage	PPL Susquehanna LLC	NRC	680	MTHM
	Susquehanna 1 & 2	Nuclear Power Plant	Wet Storage	PPL Susquehanna LLC	NRC	690	MTHM
	Three Mile Island 1	Nuclear Power Plant	Wet Storage	Exelon Generation Co.	NRC	560	MTHM
	Catawba	ISFSI	Dry Storage	Duke Power Co.	NRC	170	MTHM
	Catawba 1 & 2	Nuclear Power Plant	Wet Storage	Duke Power Co.	NRC	1050	MTHM
	Oconee	ISFSI	Dry Storage	Duke Power Co.	NRC	1240	MTHM
South Carolina	Oconee 1, 2 & 3	Nuclear Power Plant	Wet Storage	Duke Power Co.	NRC	680	MTHM
	Robinson	ISFSI	Dry Storage	Progress Energy	NRC	110	MTHM
	Robinson 2	Nuclear Power Plant	Wet Storage	Progress Energy	NRC	200	MTHM
	Summer	Nuclear Power Plant	Wet Storage	South Carolina Electric & Gas Co.	NRC	480	MTHM
	Sequoyah	ISFSI	Dry Storage	Tennessee Valley Authority	NRC	340	MTU
Tennessee	Sequoyah 1 & 2	Nuclear Power Plant	Wet Storage	Tennessee Valley Authority	NRC	830	MTHM
	Watts Bar 1	Nuclear Power Plant	Wet Storage	Tennessee Valley Authority	NRC	320	MTHM
Texas	Comanche Peak 1 & 2	Nuclear Power Plant	Wet Storage	Luminant Power	NRC	890	MTHM
	South Texas 1 & 2	Nuclear Power Plant	Wet Storage	STP Nuclear Operating Co.	NRC	1060	MTHM
Utah	Private Fuel Storage[223]	ISFSI	Dry Storage	PFS LLC	NRC	0	MTU

[223] GE Morris and the Utah Private Fuel Storage are not located at a nuclear power plant site. Private fuel storage is licensed, but not constructed.

U.S. Fourth National Report-Joint Convention on the Safety of Spent Fuel Management and on the Safety of Radioactive Waste Management

Annex D-1D Spent Fuel Management Facilities On Site Storage at Nuclear Power Plants (Inventory as of December 31, 2010, per DOE projections)[221]

State	Installation	Facility	Function	Licensee	Regulator	Inventory	Units
Vermont	Vermont Yankee	ISFSI	Dry Storage	Entergy Nuclear Operations	NRC	60	MTHM
	Vermont Yankee	Nuclear Power Plant	Wet Storage	Entergy Nuclear Operations	NRC	550	MTHM
Virginia	North Anna	ISFSI	Dry Storage	Dominion Generation	NRC	550	MTHM
	North Anna 1 & 2	Nuclear Power Plant	Wet Storage	Dominion Generation	NRC	650	MTHM
	Surry	ISFSI	Dry Storage	Dominion Generation	NRC	850	MTHM
	Surry 1 & 2	Nuclear Power Plant	Wet Storage	Dominion Generation	NRC	350	MTHM
Washington	Columbia	Nuclear Power Plant	Wet Storage	Energy Northwest	NRC	270	MTHM
	Columbia Gen. Station	ISFSI	Dry Storage	Energy Northwest	NRC	320	MTHM
Wisconsin	Kewaunee	ISFSI	Dry Storage	Dominion Generation	NRC	50	MTHM
	Kewaunee	Nuclear Power Plant	Wet Storage	Dominion Generation	NRC	410	MTHM
	LaCrosse (shutdown)	Nuclear Power Plant	Wet Storage	Dairyland Power Cooperative	NRC	40	MTHM
	Point Beach	ISFSI	Dry Storage	FPL Group (including Florida Power & Light Co., NextEra Energy Resources)	NRC	320	MTHM
	Point Beach 1 & 2	Nuclear Power Plant	Wet Storage	FPL Group (including Florida Power & Light Co., NextEra Energy Resources)	NRC	490	MTHM

Annex D-2A Radioactive Waste Management Facilities Government Facilities[224]

State	Installation	Licensee	Regulator	Facility	Function	Waste Source	Waste/ Material Type	Inventory (m³)[225]	Estimated Activity (Bq)[226]	Rad Cat
California	Lawrence Berkeley National Lab	DOE	DOE	Various Waste Facilities	Storage	2	LLW/MLLW	3.43E+03		6
				TRU Waste Facilities	Storage	2	TRU	1.10E+00	3.08E+10	3
	Lawrence Livermore National Lab	DOE	DOE	Various Waste Facilities	Storage	1	LLW/MLLW	2.42E+03		1,2,3, 4,5
						1	TRU	2.56E+02	1.05E+14	3
	Stanford Linear Accelerator	DOE	DOE	Various Waste Facilities	Storage	2	LLW/MLLW	3.56E+03		1
Idaho	Idaho Site	DOE	DOE/ID	HLW Tank Farm	Liquid Storage in underground tanks	1	HLW	3.41E+03	1.85E+16	2,3
			DOE	Calcined Solids Storage Facility	Storage in underground tanks/bins	1	HLW	4.40E+03	1.20E+18	2,3
	Idaho Site	DOE	DOE/ID/EPA	Idaho CERCLA Disposal Facility	Disposal in engineered surface disposal cell for D&D wastes	1	LLW	2.86E+05		1,2,3, 4,5
			DOE	RWMC (Includes remote RH vaults)	Disposal in shallow land disposal facility	1	LLW	6.61E+04	2.71E+17	1,2,3, 4,5
	Idaho Site	DOE	DOE/ID	Various Waste Facilities	Storage, characterization, treatment, packaging	1	LLW/MLLW	7.14E+03	1.80E+11	1,2,3, 4,5
	Idaho Site	DOE	DOE	TRU Waste Storage Facilities	Storage	1	TRU	3.23E+04	7.00E+15	2,3

224 See Key to Annex D-2 on last page of this table.
225 Stored inventories for LLW/MLLW are as of 9/30/2010 per the DOE FY2011 BLDD. Stored inventories for TRU are as of 12/31/2009 per the Annual Transuranic Waste Inventory Report - 2010. Disposed inventories for LLW/MLLW in active facilities are as of 9/30/2010 per the DOE FY2011 BLDD. Disposed TRU inventory (WIPP) is as of 12/31/2010.
226 Estimated activities for TRU stored inventories are as of 12/31/2009 per the Annual Transuranic Waste Inventory Report - 2010. Estimated activity for disposed TRU at WIPP is as of 12/31/10 per the WIPP Waste Information System. Activities for other waste types are parametric estimates.

U.S. Fourth National Report–Joint Convention on the Safety of Spent Fuel Management and on the Safety of Radioactive Waste Management

Annex D-2A Radioactive Waste Management Facilities Government Facilities[224]

State	Installation	Licensee	Regulator	Facility	Function	Waste Source	Waste/Material Type	Inventory (m³)[225]	Estimated Activity (Bq)[226]	Rad Cat
	Idaho Site	DOE	DOE/ID	AMWTP	Characterization, treatment, and packaging	1	TRU	0.00E+00		
Illinois	Argonne National Laboratory	DOE	DOE	Various Waste Facilities	Storage	2	LLW/MLLW	5.49E+03		1,2,3,4,5
				TRU Storage	Storage	1	TRU	1.18E+02	1.14E+13	2,3
	Fermi Lab	DOE	DOE	Various Waste Facilities	Storage	2	LLW/MLLW	5.83E+02		1
Kentucky	Paducah Gaseous Diffusion Plant	DOE	DOE/KY	Various Waste Facilities	Storage, characterization, treatment, packaging	1	LLW/MLLW	3.69E+02		3,4
				TRU Storage	Storage	1	TRU	6.80E+00	1.18E+11	3,4
Missouri	Weldon Springs	DOE	DOE	On-Site Disposal Cell (Closed)	Disposal in engineered, surface disposal cell (Closed)	1	AEA Section 11e.(2)	1.12E+06		4
Multiple	Other DOE	DOE	DOE/various states	Various Waste Facilities	Storage, characterization, treatment, packaging	1	LLW/MLLW	6.60E+01		6
Multiple [227]	Other DOE TRU Small Quantity Sites	DOE	DOE	TRU Waste Facilities (small)	Storage	1	TRU	1.87E+01	4.74E+13	3
Nevada	Nevada National Security Site	DOE	DOE	Area 3/Area 5 RWMS	Disposal in trenches and subsidence craters	1	LLW	1.10E+06	5.20E+17	1,2,3,4,5
	Nevada National Security Site	DOE	DOE/NV	LLW Storage Facility	Storage	1	LLW	4.50E-01		1,2,3,4,5
				MW Disposal Unit	Disposal in shallow trenches	1	MLLW	1.33E+04	6.10E+13	1,2,3,4,5

227 This entry includes multiple facilities with small inventories of LLW/MLLW stored inventories at Ames Laboratory, CA; Energy Technology Engineering Center, CA; , Kansas City Plant, Missouri; Pacific Northwest National Laboratory, WA; Thomas Jefferson National Accelerator Facility, VA;, and Waste Isolation Pilot Plant, NM; Multiple includes TRU waste inventories at General Electric Vallecitos, CA; Nuclear Radiation Development Site, Inc., NY; U.S. Army Materiel Command, IL.

158

U.S. Fourth National Report-Joint Convention on the Safety of Spent Fuel Management and on the Safety of Radioactive Waste Management

Annex D-2A Radioactive Waste Management Facilities Government Facilities[224]

State	Installation	Licensee	Regulator	Facility	Function	Waste Source	Waste/Material Type	Inventory (m³)[225]	Estimated Activity (Bq)[226]	Rad Cat
	Nevada National Security Site	DOE	DOE	Greater Confinement Disposal	Disposal in boreholes	1	TRU	2.00E+02	2.11E+15	1,2,3,4,5
				TRU Waste Facilities	Storage, characterization, packaging	1	TRU	3.91E+01	1.53E+13	3
	Los Alamos National Lab	DOE	DOE	Technical Area 54/Area G	Disposal in shallow land disposal facility	1	LLW	2.49E+05	8.40E+16	1,2,3,4,5
	Los Alamos National Lab	DOE	DOE/NM	Various Waste Facilities	Storage	1	LLW/MLLW	1.14E+03		1,2,3,4,5
	Los Alamos National Lab	DOE	DOE	Sealed Source Facilities	Disused Sealed Source Storage	1	SRC	3.33E+02 Containers	2.90E+14	6
	Los Alamos National Lab			TRU Waste Facilities	Storage, characterization, packaging	1	TRU	1.08E+04	1.18E+16	2,3
New Mexico	Sandia National Lab - NM	DOE	DOE/NM	Various Waste Facilities	Storage, characterization, treatment, packaging	1	LLW/MLLW	3.04E+01	2.20E+11	2,3,5
	Sandia National Lab - NM	DOE	DOE	TRU Waste Facilities	Storage, characterization, packaging	1	TRU	2.00E+01	2.85E+13	3
	Waste Isolation Pilot Plant	DOE	DOE/NM EPA	WIPP Disposal	Disposal in deep salt formation	1	TRU	7.24E+04	7.19E+16	1,2,3,4,5
	Niagara Falls Storage Site (FUSRAP)	USACE	NY	Niagara Falls Storage Facility	Restoration Waste Storage	1	AEA Section 11e.(2)	1.99E+05		4
New York	Separations Process Research Unit	DOE	DOE/NY	Various Waste Facilities	Storage	1	LLW/MLLW	1.78E+04		2,3
	West Valley Demonstration Plant	DOE	DOE	HLW Glass Storage Cell	Interim storage of Vitrified HLW in a former process cell	3	HLW	2.29E+02		2,3

Annex D-2A Radioactive Waste Management Facilities Government Facilities[224]

State	Installation	Licensee	Regulator	Facility	Function	Waste Source	Waste/ Material Type	Inventory (m³)[225]	Estimated Activity (Bq)[226]	Rad Cat
	West Valley Demonstration Plant	DOE	DOE/NY	Various Waste Facilities	Storage, characterization, treatment, packaging	3	LLW/MLLW	4.28E+03		1,2,3
				TRU Waste Facilities	Storage	3	TRU	1.35E+03	1.04E+16	2,3
Ohio	Fernald Environmental Management Project	DOE	DOE	On-Site Disposal Facility	Disposal (from D&D) in engineered surface disposal cell (Closed)	1	LLW	2.29E+06		4
	Portsmouth Gaseous Diffusion Plant	DOE	DOE/OH	Various Waste Facilities	Storage, treatment, packaging	1	LLW/MLLW	5.72E+03		4
				Saltstone Vaults	Disposal of low-activity fraction of HLW	1	LLW	5.31E+04		1,2,3, 4,5
				HLW Tank Farm	Liquid Storage in underground double-shell, stainless steel tanks	1	HLW	1.25E+05		1,2,3, 4,5
South Carolina	Savannah River Site	DOE	DOE	Glass Waste Storage Building	Interim Storage of Vitrified HLW[5]	1	HLW	2.38E+03		1,2,3, 4,5
				Defense Waste Processing Fac.	HLW Liquid Treatment (Vitrification)	1	HLW	0.00E+00		
				E-Area Disposal	Disposal in underground vaults and trenches	1	LLW	3.20E+05	4.80E+16	1,2,3, 4,5
				Old Burial Ground	Historic disposal (Closed)	1	LLW	6.77E+05		1,2,3, 4,5

U.S. Fourth National Report-Joint Convention on the Safety of Spent Fuel Management and on the Safety of Radioactive Waste Management

Annex D-2A Radioactive Waste Management Facilities Government Facilities[224]

State	Installation	Licensee	Regulator	Facility	Function	Waste Source	Waste/ Material Type	Inventory (m³)[225]	Estimated Activity (Bq)[226]	Rad Cat
	Savannah River Site	DOE	DOE/SC	Various Waste Facilities	Storage, characterization, treatment, packaging	1	LLW/MLLW	2.13E+03		1,2,3,4,5
	Savannah River Site	DOE	DOE	TRU Waste Facilities	Storage, characterization, packaging	1	TRU	3.61E+03	8.28E+15	2,3
			DOE	Hydrofracture	Historic disposal (Closed)	1	LLW	1.73E+04		1,2,3,4,5
				Old Burial Ground	Historic disposal (Closed)	1	LLW	4.41E+05		1,2,3,4,5
				Interim Waste Management Fac.	Disposal in engineered aboveground facility (Closed)	1	LLW	3.70E+03	1.18E+13	1,2,3,4,5
Tennessee	Oak Ridge Reservation	DOE	DOE/TN EPA	EMWMF	Disposal in engineered surface disposal cell for D&D wastes	1	LLW	9.49E+05		1,2,3,4,5
			DOE/TN	Various Waste Facilities	Storage (in building and on concrete pad), characterization, treatment, packaging	1	LLW/MLLW	3.19E+03		1,2,3,4,5
			DOE	TRU Waste Facilities	Storage, characterization, packaging, treatment	1	TRU	1.66E+03	2.97E+14	2,3
Utah	Monticello Remedial Action Project	DOE	DOE	Monticello Disposal Cell	Disposal in engineered, surface disposal cell (Closed)	1	AEA Section 11e.(2)	1.91E+06		4
Washington	Hanford Site	DOE	DOE/WA	HLW Tank Farm	Liquid Storage in underground single-and double-shell tanks	1	HLW	2.04E+05		1,2,3,4,5

U.S. Fourth National Report-Joint Convention on the Safety of Spent Fuel Management and on the Safety of Radioactive Waste Management

Annex D-2A Radioactive Waste Management Facilities Government Facilities[224]

State	Installation	Licensee	Regulator	Facility	Function	Waste Source	Waste/Material Type	Inventory (m³)[225]	Estimated Activity (Bq)[226]	Rad Cat
			DOE	200 Area Burial Grounds	Disposal in trenches	1	LLW	3.10E+05	1.68E+17	1,2,3, 4,5
		DOE	DOE/WA/ EPA	ERDF	Disposal (from D&D) in engineered surface disposal unit	1	LLW	4.97E+06		1,2,3, 4,5
	Hanford Site			Decommissioned Submarine Hulls Disposal Area	Navy submarine hulls disposal in trenches	1	LLW	1.22E+02		1
	Hanford Site	DOE	DOE/WA	IDF	Disposal	1	LLW/MLLW	0.00E+00		1,2,3, 4,5
				Various Waste Facilities	Storage, characterization, treatment, packaging	1	LLW/MLLW	1.91E+02		1,2,3, 4,5
				RMW Trenches	Disposal in lined trenches	1	MLLW	1.12E+05	6.00E+16	1,2,3, 4,5
	Hanford Site	DOE	DOE	WESF	Cs-Sr Storage in hot cells and storage pool	1	SRC	1.93E+03 Sources	2.85E+18	2
				TRU Waste Facilities	Storage, characterization, packaging	1	TRU	2.12E+04	5.21E+16	2,3

Annex D-2 Key

Waste Source		Category	Radionuclide Category — Key Isotopes
Defense applications	1	Activation Products	Primarily ^{36}Cl, ^{55}Fe, ^{54}Mn, ^{65}Zn, ^{58}Co, ^{60}Co, ^{63}Ni,
Nuclear applications	2	Mixed Fission Products	Radioactive isotopes and daughters from ^{72}Zn to ^{158}Gd; primary longer-lived isotopes are ^{85}Kr, ^{89}Sr, $^{90/Y-90}$Sr, ^{91}Y, ^{95}Zr, ^{95}Nb, $^{103/Rh103}$Ru, $^{106/Rh-106}$Ru, $^{125/Te-125}$Sb, $^{137/Ba-137}$Cs, ^{141}Ce, $^{144/Pr-144}$Ce, ^{147}Pm, ^{m151}Pm, ^{m151}S, and ^{155}Eu
Commercial	3	Transuranic Isotopes	Isotopes of Cf, Bk, Cm, Am, Pu, and Np, and their respective decay products.
	4	Naturally-Occurring Isotopes	^{238}U, ^{235}U, ^{234}U, ^{232}Th, and their respective decay products (^{231}Pa, ^{227}Th, ^{228}Th, ^{230}Th, ^{231}Th, ^{234}Th, ^{227}Ac, ^{228}Ac, ^{223}Ra, ^{224}Ra, ^{226}Ra, ^{228}Ra, ^{223}Fr, ^{219}Rn, ^{220}Rn, ^{222}Rn, ^{215}At, ^{218}At, ^{219}At, ^{210}Po, ^{211}Po, ^{212}Po, ^{214}Po, ^{215}Po, ^{216}Po, ^{218}Po, ^{210}Bi, ^{211}Bi, ^{212}Bi, ^{214}Bi, ^{147}Sm, ^{148}Sm, ^{149}Sm, ^{14}C, ^{40}K, ^{40}V, ^{87}Rb, ^{115}In, ^{123}Te, ^{138}La, ^{142}Ce, ^{144}Nd, ^{210}Pb, ^{211}Pb, ^{212}Pb, ^{214}Pb, ^{206}Tl, ^{207}Tl, ^{208}Tl, and ^{210}Tl), ^{152}Gd, ^{156}Dy, ^{176}Lu, ^{174}Hf, ^{180}Ta, ^{187}Re, ^{190}Pt, ^{204}Pb, ^{215}Bi
	5	Tritium	^{3}H
	6	Various	Radioactivity from various sources and categories

Annex D-2B Radioactive Waste Management Facilities Commercial/Other Facilities [228]

State	Installation	Licensee	Regulator	Type	Function	Waste Source	Inventory (m³) [229]	Estimated Activity (Bq) [230]	Rad Cat
	Advanced Chemical Transport, Inc.	Advanced Chemical Transport, Inc.	CA	Commercial	Repackage/Broker/Transporter	3	NA		
	B&B Environmental Safety	B&B Environmental Safety	CA	Commercial	Broker/Transporter	3	NA		
	Environmental Management & Controls, Inc	Environmental Management & Controls, Inc	CA	Commercial	Broker/Transporter	3	NA		
California	Environmental Management Services, Inc.	Environmental Management Services, Inc.	CA	Commercial	Broker/Transporter	3	NA		
	New World Technology	New World Technology	CA	Commercial	Processor – Waste Treatment Service (Other than compaction)	3	NA		
	PWN Environmental	PWN Environmental	CA	Commercial	Broker/Processor/Transporter	3	NA		
	Thomas Grey Associates	Thomas Grey Associates	CA	Commercial	Processor – Processing of liquids and radium	3	NA		
	Cabrera Services, Inc.	Cabrera Services, Inc.	NRC	Commercial	Processor – Decontamination Services	3	NA		
Connecticut	Radiation Safety Associates	Radiation Safety Associates	NRC	Commercial	Processor	3	NA		
	Yale Univ. Radiation Safety Section	Yale University	NRC	Academic	Processor – Academic Type A Broad	3	NA		
Florida	Perma-Fix of Florida, Inc.	Perma-Fix	FL	Commercial	Processor	3	NA		
Idaho	Qal-Tek Associates	Qal-Tek Associates	NRC	Commercial	Packaging/Broker (Sealed Sources)	3	NA		

[228] See Key to Annex D-2 on last page of this table.
[229] Commercial disposal inventories are as of 12/31/10 based on the Integrated Data Base Report (DOE/RW-0006, Rev. 13) and the Manifest Information Management System (MIMS). GTCC stored inventories based on ANL/EVS/R-10/1 (West Valley stored GTCC-like waste is reported as TRU - see Annex D-2A)
[230] Estimated activities associated with commercial disposed waste are as of 12/31/10 based on the Integrated Data Base Report (DOE/RW-0006, Rev. 13) and MIMS.

Annex D-2B Radioactive Waste Management Facilities Commercial/Other Facilities [228]

State	Installation	Licensee	Regulator	Type	Function	Waste Source	Inventory (m³) [229]	Estimated Activity (Bq) [230]	Rad Cat
Illinois	ADCO Services Inc.	ADCO Services Inc.	IL	Commercial	Processor – Processing of uranium and thorium	3	NA		
	Dept. Of The Army Rock Island Arsenal	Dept. Of The Army	NRC	Other Government	Processor/Storage – Waste Disposal Service Processing and/or Repackaging.	3	NA		
	Sheffield	State of Illinois	IL	Commercial	LLW– All Classes (Closed)	3	8.83E+04	2.23E+15	1,2,3, 4,5
Kentucky	Maxey Flats	State of Kentucky	KY	Commercial	LLW– All Classes (Closed)	3	1.35E+05	8.88E+16	1,2,3, 4,5
Maryland	Chesapeake Nuclear Services, Inc.	Chesapeake Nuclear Services, Inc.	MD	Commercial	Broker	3	NA		
	Clym Environmental Services, LLC	Clym Environmental Services, LLC	MD	Commercial	Broker	3	NA		
	Dept. Of The Army Ft. Detrick	Dept. Of The Army	NRC	Other Government	Processor – Waste Disposal Service Processing and/or Repackaging.	3	NA		
	Ecology Services	Ecology Services	MD	Commercial	Processor – Mixed waste processing	3	NA		
	RSO, Inc.	RSO, Inc.	MD	Commercial	Broker/Transporter	3	NA		
Michigan	Pharmacia & Upjohn Company	Pharmacia & Upjohn Company	NRC	Private	Processor – Manufacturing and Distribution Type A Broad	3	NA		
Minnesota	University of Minnesota	University of Minnesota	MN	Academic	Processor – Waste Disposal Service Processing and/or Repackaging.	3	NA		
	Pharmacia Corporation	Pharmacia Corporation	NRC	Commercial	Processor – Waste Disposal Service Processing and/or Repackaging.	3	NA		
Missouri	R.M. Wester	R.M. Wester	NRC	Commercial	Processor	3	NA		
	R&R Trucking	R&R Trucking	NRC	Commercial	Broker/Transporter	3	NA		
	Westinghouse Electric Company, LLC	Westinghouse Electric Company, LLC	NRC	Commercial	Processor – Decommissioning of Uranium Fuel Fabrication Plants	3	NA		

Annex D-2B Radioactive Waste Management Facilities Commercial/Other Facilities [228]

State	Installation	Licensee	Regulator	Type	Function	Waste Source	Inventory (m³) [229]	Estimated Activity (Bq) [230]	Rad Cat
Montana	HHS, Dept. Of USPHS, NIH, Rocky Mountain Laboratories	Dept. of Health & Human Services	NRC	Other Government	Processor – Research and Development Type A Broad	3	NA		
Nevada	Beatty	None	NV	Commercial	LLW– All Classes (Closed)	3	1.37E+05	2.37E+16	1,2,3, 4,5
New Jersey	BASF Corporation	BASF Corporation	NRC	Commercial	Processor – Research and Development Type A Broad	3	NA		
New Jersey	Radiation Science, Inc.	Radiation Science, Inc.	NRC	Commercial	Processor – Waste Disposal Service Prepackaged only.	3	NA		
New Jersey	Teledyne Brown Engineering, Inc.	Teledyne Brown Engineering, Inc.	NRC	Commercial	Processor – Waste Disposal Service Prepackaged only.	3	NA		
New Jersey	Radiac Research Corp.	Radiac Research Corp.	NY	Commercial	Processor – Waste Disposal Service Prepackaged only.	3	NA		
New York	West Valley NRC-licensed Disposal Area and State-licensed Disposal Area	New York State Energy Research and Development Administration	NRC & NY	Commercial	LLW– All Classes (Closed)	3	7.71E+04	3.09 E +16	
North Carolina	HHS, Dept. Of Public Health Service	Dept. of Health & Human Services	NRC	Other Government	Processor – Research and Development Type A Broad	3	NA		
North Carolina	V.A. Medical Center	Dept. of Veterans Affairs	NRC	Other Government	Processor – Medical Institution Broad	3	NA		
Ohio	Solutient Technologies	Solutient Technologies	OH	Commercial	Processor – Processing	3	NA		

Annex D-2B Radioactive Waste Management Facilities Commercial/Other Facilities [231]

State	Installation	Licensee	Regulator	Type	Function	Waste Source	Inventory (m³) [232]	Estimated Activity (Bq) [233]	Rad Cat
	Alaron Corporation	Alaron Corporation	PA	Commercial	Processor – Waste Disposal Service Processing and/or Repackaging.	3	NA		
	Applied Health Physics, Inc.	Applied Health Physics, Inc.	PA	Commercial	Processor – Waste Disposal Service Prepackaged only.	3	NA		
	B&W Nuclear Environmental Services	BWX Technologies	NRC	Commercial	Processor – Decommissioning of Advanced Fuel R&D and Pilot Plants	3	NA		
Pennsylvania	BWX Technologies, Inc.	BWX Technologies	NRC	Commercial	Processor – Decommissioning of Advanced Fuel R&D and Pilot Plants	3	NA		
	Fox Chase Cancer Center	Fox Chase Cancer Center	PA	Private	Processor – Medical Institution Broad	3	NA		
	MHF Logistical Solutions, Inc.	MHF Logistical Solutions, Inc.	PA	Commercial	Broker/Transporter	3	NA		
	Energy Solutions (Barnwell)	Energy Solutions (Barnwell)	SC	Commercial	LLW Disposal Class A	3	7.18E+05	7.57E+17	1,2,3, 4,5
South Carolina			SC	Commercial	LLW Disposal Class B	3	5.29E+04		
					LLW Disposal Class C	3	2.68E+04		
	GTS-Duratek/ Chem-Nuclear Systems, Inc.	GTS-Duratek, Inc	SC	Commercial	Processor – Decommissioning of Byproduct Material Facilities	3	NA		
	Hittman Transport Services	EnergySolutions	SC	Commercial	Transporter	3	NA		
Tennessee	Ameriphysics, LLC	Ameriphysics, LLC	TN	Commercial	Broker	3	NA		

[231] See Key to Annex D-2 on last page of this table.
[232] Commercial disposal inventories are as of 12/31/10 based on the Integrated Data Base Report (DOE/RW-0006, Rev. 13) and the Manifest Information Management System (MIMS). GTCC stored inventories based on ANL/EVS/R-10/1 (West Valley stored GTCC-like waste is reported as TRU - see Annex D-2A)
[233] Estimated activities associated with commercial disposed waste are as of 12/31/10 based on the Integrated Data Base Report (DOE/RW-0006, Rev. 13) and MIMS.

U.S. Fourth National Report-Joint Convention on the Safety of Spent Fuel Management and on the Safety of Radioactive Waste Management

Annex D-2B Radioactive Waste Management Facilities Commercial/Other Facilities [234]

State	Installation	Licensee	Regulator	Type	Function	Waste Source	Inventory (m³) [235]	Estimated Activity (Bq) [236]	Rad Cat
	Bionomics	Bionomics	TN	Commercial	Processor	3	NA		
	Chase Environmental	Chase Environmental Group, Inc.	TN	Commercial	Processor	3	NA		
	Diversified Technologies Services, Inc	Diversified Technologies Services, Inc	TN	Commercial	Processor (Liquid wastes)	3	NA		
	Energy Solutions LLC	Energy Solutions Bear Creek	TN	Commercial	Processor	3	NA		
	Energy Solutions LLC	Energy Solutions Gallagher Road	TN	Commercial	Processor	3	NA		
	Impact Services, Inc.	Impact Services, Inc.	TN	Commercial	Processor	3	NA		
Tennessee	Perma-Fix DSSI	Perma-Fix	TN	Commercial	Processing of resins, sludges, and liquids	3	NA		
	Perma-Fix M&EC	Perma-Fix	TN	Commercial	Processing/treatment of mixed wastes	3	NA		
	Philotechnics	Philotechnics, Ltd.	TN	Commercial	Broker/Processor	3	NA		
	Studsvik (formerly RACE, LLC)	Studsvik Processing Facility Memphis, LLC	TN	Commercial	Processor – Processing of large equipment	3	NA		
	Studsvik	Studsvik Processing Facility Erwing, TN	TN	Commercial	Processing Treatment	3	NA		
	Toxco Incorporated	Toxco Incorporated	TN	Commercial	Processor	3	NA		
	V.A. Medical Center	Dept. of Veterans Affairs	NRC	Other Government	Processor – High Dose Rate Remote Afterloader	3	NA		
Texas	MKM Engineers, Inc.	MKM Engineers, Inc.	TX	Commercial	Processor – Waste Disposal Service Processing and/or Repackaging.	3	NA		

[234] See Key to Annex D-2 on last page of this table.
[235] Commercial disposal inventories are as of 12/31/10 based on the Integrated Data Base Report (DOE/RW-0006, Rev. 13) and the Manifest Information Management System (MIMS). GTCC stored inventories based on ANL/EVS/R-10/1 (West Valley stored GTCC-like waste is reported as TRU - see Annex D-2A)
[236] Estimated activities associated with commercial disposed waste are as of 12/31/10 based on the Integrated Data Base Report (DOE/RW-0006, Rev. 13) and MIMS.

168

U.S. Fourth National Report-Joint Convention on the Safety of Spent Fuel Management and on the Safety of Radioactive Waste Management

Annex D-2B Radioactive Waste Management Facilities Commercial/Other Facilities[237]

State	Installation	Licensee	Regulator	Type	Function	Waste Source	Inventory (m³)[238]	Estimated Activity (Bq)[239]	Rad Cat
	NSSI	NSSI	TX	Commercial	MLLW processing	3	NA		
	Specpro, Inc.	Specpro, Inc.	TX	Commercial	Processor – Waste Disposal Service Processing and/or Repackaging.	3	NA		
	USA Environment	USA Environment	TX	Commercial	Broker/Processor/Transporter	3	NA		
Texas					GTCC Storage	2	4.30E+01		1,3
					MLLW Treatment	3	NA		
					11e.(2) Storage and Disposal	1	2.12E+04		4
	Waste Control Specialists (WCS)	WCS	TX	Commercial	Commercial LLW Disposal (Under Construction)	3	0.00E+00		
					Federal LLW Disposal (Under Construction)	1,2	0.00E+00		
Utah	EnergySolutions	EnergySolutions	UT	Commercial	MLLW Treatment and Disposal	3	1.11E+05		1,2,3,4,5
					LLW–Class A Disposal	3	3.11E+06	1.42E+15	1,2,3,4,5
					11e.(2) Disposal	3	1.35E+06		1,2,3,4,5
Virginia	B&W Energy Services	B&W Energy Services	NRC	Commercial	TRU Storage (DOE)	1	1.00E+01	7.81E+13	3

[237] See Key to Annex D-2 on last page of this table.
[238] Commercial disposal inventories are as of 12/31/10 based on the Integrated Data Base Report (DOE/RW-0006, Rev. 13) and the Manifest Information Management System (MIMS). GTCC stored inventories based on ANL/EVS/R-10/1 (West Valley stored GTCC-like waste is reported as TRU - see Annex D-2A)
[239] Estimated activities associated with commercial disposed waste are as of 12/31/10 based on the Integrated Data Base Report (DOE/RW-0006, Rev. 13) and MIMS.

U.S. Fourth National Report-Joint Convention on the Safety of Spent Fuel Management and on the Safety of Radioactive Waste Management

Annex D-2B Radioactive Waste Management Facilities Commercial/Other Facilities[240]

State	Installation	Licensee	Regulator	Type	Function	Waste Source	Inventory (m³)[241]	Estimated Activity (Bq)[242]	Rad Cat
	PermaFix Northwest	PermaFix Northwest	WA	Commercial	MLLW treatment and processing	3	NA		
Washington	US Ecology - Richland	US Ecology	WA	Commercial	LLW–Class A Disposal	3	3.90E+05	2.73E+17	1,2,3, 4,5
					LLW–Class B Disposal	3	3.72E+03		
					LLW–Class C Disposal	3	2.79E+03		
	Covance Laboratories	Covance Laboratories	WI	Private	Processor– Research and Development Other	3	NA		
Wisconsin	William S. Middleton Memorial V.A. Hospital	Dept. of Veterans Affairs	NRC	Other Government	Processor– Medical Institution Broad	3	NA		
Multiple	Multiple ISFSIs	Various utilities	NRC	Commercial	GTCC Storage	3	8.70E+01		1,3
Past Practices									
Ocean Disposal	Atlantic			Past Practice	LLW	1,2,3	8.60E+03	2.94E+15	
	Pacific			Past Practice	LLW	1,2,3	1.40E+04	5.54E+14	

[240] See Key to Annex D-2 on last page of this table.
[241] Commercial disposal inventories are as of 12/31/10 based on the Integrated Data Base Report (DOE/RW-0006, Rev. 13) and the Manifest Information Management System (MIMS). GTCC stored inventories based on ANL/EVS/R-10/1 (West Valley stored GTCC-like waste is reported as TRU - see Annex D-2A)
[242] Estimated activities associated with commercial disposed waste are as of 12/31/10 based on the Integrated Data Base Report (DOE/RW-0006, Rev. 13) and MIMS.

Annex D-2 Key

	Waste Source		Category	Radionuclide Category — Key Isotopes
		1	Defense applications	Activation Products	Primarily ^{36}Cl, ^{55}Fe, ^{54}Mn, ^{65}Zn, ^{58}Co, ^{60}Co, ^{63}Ni;
		2	Nuclear applications	Mixed Fission Products	Radioactive isotopes and daughters from ^{72}Zn to ^{158}Gd; primary longer-lived isotopes are ^{85}Kr, ^{89}Sr, $^{90/Y\cdot90}$Sr, ^{91}Y, ^{95}Zr, ^{95}Nb, $^{103/Rh103}$Ru, $^{106/Rh\cdot106}$Ru, $^{125Te\cdot125}$Sb, $^{137/Ba\cdot137}$Cs, ^{141}Ce, $^{144/Pr\cdot144}$Ce, ^{147}Pm, ^{m151}S, and ^{155}Eu
		3	Commercial	Transuranic Isotopes	Isotopes of Cf, Bk, Cm, Am, Pu, and Np, and their respective decay products.
		4		Naturally-Occurring Isotopes	^{238}U, ^{235}U, ^{234}U, ^{232}Th, and their respective decay products (^{231}Pa, ^{227}Th, ^{228}Th, ^{230}Th, ^{231}Th, ^{234}Th, ^{227}Ac, ^{228}Ac, ^{223}Ra, ^{224}Ra, ^{226}Ra, ^{228}Ra, ^{223}Fr, ^{219}Rn, ^{220}Rn, ^{222}Rn, ^{215}At, ^{218}At, ^{210}Po, ^{211}Po, ^{212}Po, ^{214}Po, ^{215}Po, ^{216}Po, ^{218}Po, ^{210}Bi, ^{211}Bi, ^{212}Bi, ^{214}Bi, ^{210}Pb, ^{211}Pb, ^{212}Pb, ^{214}Pb, ^{206}Tl, ^{207}Tl, ^{208}Tl, and ^{210}Tl), ^{14}C, ^{40}K, ^{40}V, ^{87}Rb, ^{115}In, ^{123}Te, ^{138}La, ^{142}Ce, ^{144}Nd, ^{147}Sm, ^{148}Sm, ^{149}Sm, ^{152}Gd, ^{156}Dy, ^{176}Lu, ^{174}Hf, ^{180}Ta, ^{187}Re, ^{190}Pt, ^{204}Pb, ^{215}Bi
		5		Tritium	^{3}H
		6		Various	Radioactivity from various sources and categories

Annex D-3A Uranium Mill Tailings and Related Sites

State	Site Name/Location	Licensee	Type	Status	Regulator	Regulatory Program	Quantity of Contaminated Material	Quantity Units	Total ^{226}Ra-Activity (TBq)
Arizona	Tuba City	DOE	Surface residual radioactive material disposal cell	Under general NRC license, in DOE LTSP program; property owned by Navajo Indian Nation.	NRC	UMTRCA Title I	2250000	Dry Tonnes	35
	Cheney Disposal Cell (residual radioactive material removed from the former Grand Junction Climax site)	DOE	Surface residual radioactive material disposal cell	Active until 2023 to accept residual radioactive material from other sites.	NRC	UMTRCA Title I	3597149	m³	TBD
	Cotter	Cotter Corp.-USA	Conventional mill	Standby/periodic limited operations	Colorado	UMTRCA Title II	2000000	Dry Tonnes	N/A
	Cotter Schwarzwalder Mine	Cotter Corp.-USA	Uranium mine ore separation and size reduction	Physical completion planned for 2009/Ore stockpiling under consideration.	Colorado	Colorado Mined Land Reclamation Act	6150	Dry Tonnes	N/A
	Durita	Hecla Mining Company	Heap Leach Site	Physical completion 1998. Reclamation/Stability monitoring	Colorado	UMTRCA Title II	540000	Dry Tonnes	N/A
Colorado	Durango	DOE	Conventional mill and surface residual radioactive material disposal cell	Under general NRC license, in DOE LTSP program	NRC	UMTRCA Title I	3700000	Dry Tonnes	52
	Gunnison	DOE	Conventional mill and surface residual radioactive material disposal cell	Under general NRC license, in DOE LTSP program	NRC	UMTRCA Title I	1140000	Dry Tonnes	6.5
	Homestake Mining and Pitch	Homestake Mining Company	Mine drainage treatment & residuals repository	US Forest Service land; joint regulation with Colorado Division of Reclamation Mining & Safety	US Forest Service & Colorado	UMTRCA Title II	N/A		N/A

Annex D-3A Uranium Mill Tailings and Related Sites

State	Site Name/Location	Licensee	Type	Status	Regulator	Regulatory Program	Quantity of Contaminated Material	Quantity Units	Total ^{226}Ra- Activity (TBq)
	Maybell	DOE	Conventional mill and surface residual radioactive material disposal cell	Under general NRC license, in DOE LTSP program; annual groundwater monitoring inspections.	NRC	UMTRCA Title I	4291928	Dry Tonnes	17
	Maybell - West	DOE	Heap Leach Site	Decommissioning determination pending. Site is closed.	Colorado	UMTRCA Title II	1800000	Dry Tonnes	3.6
	Naturita	DOE	Surface residual radioactive material disposal cell	Under general NRC license, in DOE LTSP program.	NRC	UMTRCA Title I	971762	Dry Tonnes	2.9
	Rifle	DOE	Surface residual radioactive material disposal cell	Under general NRC license, in DOE LTSP program	NRC	UMTRCA Title I	4967451	Dry Tonnes	101
	Slick Rock	DOE	Surface residual radioactive material disposal cell	Under general NRC license, in DOE LTSP program	NRC	UMTRCA Title I	1140000	Dry Tonnes	6.5
	Sweeney	EPA Superfund	Conventional mill	No activity; site under State Order	Colorado	UMTRCA Title II	N/A		N/A
	UMETCO/ Uravan	EPA Superfund	Conventional mill	Reclamation/decommissioning	Colorado	UMTRCA Title II	9500000	Dry Tonnes	N/A
Idaho	Lowman	DOE	Surface residual radioactive material disposal cell	Under general NRC license, in DOE LTSP program	NRC	UMTRCA Title I	222230	Dry Tonnes	0.4
Illinois	TRONOX (Formerly Kerr McGee) West Chicago	Kerr-McGee	Conventional thorium mill	Decommissioning. The West Chicago site is being decommissioned for unrestricted use. Anticipate finishing cleanup in 2010	Illinois	UMTRCA Title II		Materials relocated to Utah disposal site	N/A

U.S. Fourth National Report-Joint Convention on the Safety of Spent Fuel Management and on the Safety of Radioactive Waste Management

Annex D-3A Uranium Mill Tailings and Related Sites

State	Site Name/Location	Licensee	Type	Status	Regulator	Regulatory Program	Quantity of Contaminated Material	Quantity Units	Total ^{226}Ra-Activity (TBq)
Nebraska	Crow Butte	Crow Butte Resources, Inc.	In situ site	Operating	NRC	UMTRCA Title II	N/A		N/A
	Ambrosia Lake	DOE	Conventional mill and surface residual radioactive material disposal cell	Under general NRC license, in DOE LTSP program	NRC	UMTRCA Title I	6931000	Dry Tonnes	69
	Ambrosia Lake	Rio Algom Mining LLC	Conventional mill	Not yet on LTSP. Mill decommissioned in 2003, surface reclamation projected finish 2010 {possible license transfer and new mill construction}	NRC	UMTRCA Title II	30100000	Dry Tonnes	N/A
	Bluewater	DOE	Conventional mill and surface mill tailings disposal cell	Under general NRC license, in DOE LTSP program	NRC	UMTRCA Title II	24000000	Dry Tonnes	457
New Mexico	Church Rock	United Nuclear Corporation Mining and Milling	Conventional mill; groundwater restoration program	Not yet on LTSP. DP[243] approved 3/1991, groundwater restoration projected finish in 2011	NRC	UMTRCA Title II	3200000	Dry Tonnes	N/A
	Crown Point	Hydro Resources, Inc.	In situ site	Partially permitted	NRC	UMTRCA Title II	N/A		N/A
	Grants	Homestake Mining Co	Conventional mill; groundwater restoration program	Not yet on LTSP. Revised DP approved 3/1995, groundwater restoration projected finish 2017	NRC	UMTRCA Title II	20300000	Dry Tonnes	N/A
	L–Bar (Sohio Western Mining)	DOE	Conventional mill and surface mill tailings disposal cell	Under general NRC license, in DOE LTSP program	NRC	UMTRCA Title II	1900000	Dry Tonnes	N/A

243 DP= Decommissioning Plan

174

U.S. Fourth National Report-Joint Convention on the Safety of Spent Fuel Management and on the Safety of Radioactive Waste Management

Annex D-3A   Uranium Mill Tailings and Related Sites

State	Site Name/Location	Licensee	Type	Status	Regulator	Regulatory Program	Quantity of Contaminated Material	Quantity Units	Total ^{226}Ra-Activity (TBq)
	Shiprock	DOE	Conventional mill and surface residual radioactive material disposal cell	Under general NRC license, in DOE LTSP program	NRC	UMTRCA Title I	2520000	Wet Tonnes	28
Oklahoma	Sequoyah Fuels Corporation	Sequoyah Fuels Corp.	UF6 Facility	NRC reviewing the groundwater-corrective action plan; cleanup estimate completion by end of 2012.	NRC	UMTRCA Title II	248318	m^3	0.11
Oregon	Lakeview	DOE	Conventional mill and surface residual radioactive material disposal cell	Under general NRC license, in DOE LTSP program	NRC	UMTRCA Title I	736000	Dry Tonnes	1.6
	Burrell	DOE	Surface residual radioactive material disposal cell	Under general NRC license, in DOE LTSP program; groundwater monitoring and maintenance program to maintain site integrity.	NRC	UMTRCA Title I	86000	Dry Tonnes	0.15
Pennsylvania	Canonsburg	DOE	Conventional mill and surface residual radioactive material disposal cell	Under general NRC license, in DOE LTSP program; surface and groundwater under monitoring regime.	NRC	UMTRCA Title I	226000	Dry Tonnes	4
South Dakota	Edgemont	DOE	Conventional mill and surface mill tailings disposal cell	Under general NRC license, in DOE LTSP program	NRC	UMTRCA Title II	4000000	Dry Tonnes	19

U.S. Fourth National Report–Joint Convention on the Safety of Spent Fuel Management and on the Safety of Radioactive Waste Management

Annex D-3A Uranium Mill Tailings and Related Sites

State	Site Name/Location	Licensee	Type	Status	Regulator	Regulatory Program	Quantity of Contaminated Material	Quantity Units	Total ²²⁶Ra-Activity (TBq)
Texas	Uranium One Inc./ Bruni	Uranium One	In situ site	Restoration/Closure under way	Texas	UMTRCA Title II	N/A		N/A
	Conoco Conquista	Conoco Conquista	Conventional mill	All structures and equipment have been removed from site & tailings impoundment has been capped with a vegetative cover. Reclamation/Stability monitoring	Texas	UMTRCA Title II	11800000	Dry Tonnes	N/A
	Hobson, Tex-1 & Mt. Lucas Projects/ Dinero (3 locations)	South Texas Mining Venture	In situ site	Hobson is an active sites; Tex-1 & Mt. Lucas in reclamation phases.	Texas	UMTRCA Title II	N/A		N/A
	La Palangana	South Texas Mining Venture	In situ site	Under development January 2010	Texas	UMTRCA Title II	N/A		N/A
	Ray Point Felder	Exxon	Conventional mill	All structures and equipment have been removed from site & tailings impoundment has been capped with a vegetative cover. Reclamation/Stability monitoring	Texas	UMTRCA Title II	400000	Dry Tonnes	N/A
	Falls City	DOE	Conventional mill and surface residual radioactive material disposal cell	Under general NRC license, in DOE LTSP program.	NRC	UMTRCA Title I	7143000	Dry Tonnes	47
	Zamzow & Lamprech Projects /S. Texas	International Energy Corporation	In situ site	Restoration/Closure under way	Texas	UMTRCA Title II	N/A		N/A
	Alta Mesa	Mestena Uranium LLC	In situ site	Operational	Texas	UMTRCA Title II	N/A		N/A

Annex D-3A Uranium Mill Tailings and Related Sites

State	Site Name/Location	Licensee	Type	Status	Regulator	Regulatory Program	Quantity of Contaminated Material	Quantity Units	Total ^{226}Ra-Activity (TBq)
	RGR/Chevron (aka Panna Maria)	Rio Grande Resources Corporation	Conventional mill	All structures and equipment have been removed from site & tailings impoundment has been capped with a vegetative cover. Reclamation/Stability monitoring	Texas	UMTRCA Title II	5900000	Dry Tonnes	N/A
	Hobson	Rio Grande Resources Corporation	Resin Processing	Under Review	Texas	UMTRCA Title II	N/A		N/A
	Kingsville Dome	URI	In situ site	On Standby Status	Texas	UMTRCA Title II	N/A		N/A
	Vasquez	URI	In situ site	On Standby Status	Texas	UMTRCA Title II	N/A		N/A
	Andrews County	WCS	11e.(2) byproduct material disposal cell	Operational	Texas	UMTRCA Title II	N/A		N/A
	Moab/ Crescent Junction	DOE	11e.(2) byproduct material disposal site	Operational	NRC	UMTRCA Title I	3132000	Dry Tonnes	N/A
	Green River	DOE	Conventional mill and surface residual radioactive material disposal cell	Under general NRC license, in DOE LTSP program; groundwater monitoring regime.	NRC	UMTRCA Title I	501000	Dry Tonnes	1.1
Utah	Energy Solutions	Energy Solutions	11e.(2) byproduct material disposal cell	Operational	Utah	UMTRCA Title II	N/A		N/A
	Lisbon	Rio Algom Mining Corp	Conventional mill	Not yet on LTSP; reclamation completion in 2010. This is a candidate for restart.	Utah	UMTRCA Title II	3500000	Dry Tonnes	N/A
	Mexican Hat	DOE	Conventional mill and surface residual radioactive	Under general NRC license, in DOE LTSP program	NRC	UMTRCA Title I	4400000	Dry Tonnes	67

U.S. Fourth National Report-Joint Convention on the Safety of Spent Fuel Management and on the Safety of Radioactive Waste Management

Annex D-3A Uranium Mill Tailings and Related Sites

State	Site Name/Location	Licensee	Type	Status	Regulator	Regulatory Program	Quantity of Contaminated Material	Quantity Units	Total ^{226}Ra-Activity (TBq)
			material disposal cell						
	Moab	DOE	Mill & Tailings Disposal	Under active reclamation by DOE; site will not come under general license in 10 CFR 40.27 until surface reclamation is complete	NRC	UMTRCA Title I	7668000	Dry Tonnes	N/A
	Salt Lake City Disposal Cell (Clive)	DOE	Surface residual radioactive material disposal cell	Under general NRC license, in DOE LTSP program.	NRC	UMTRCA Title I	2798000	Dry Tonnes	57
	Salt Lake City Processing Site Central Valley Water Reclamation Facility	DOE	Currently a sewage treatment facility	Institutional controls maintained by DOE	DOE	UMTRCA Title I		Residual ^{226}Ra- and ^{230}Th-contaminated material	N/A
	Shootaring Canyon	Plateau Resources Ltd	Conventional uranium mill	On standby status	Utah	UMTRCA Title II	15300 m^3 – only operated for 3 months		N/A
	White Mesa	International Uranium Corporation	Conventional uranium mill	Operating, capable of processing alternate feed	Utah	UMTRCA Title II	3200000	Dry Tonnes	N/A
Washington	Dawn Mining	Dawn Mining Company	Conventional uranium mill	Reclamation/ Residue Disposal. Final reclamation December 31, 2013	WA[244]	UMTRCA Title II	2800000	Dry Tonnes	N/A
	WNI Sherwood	DOE	Conventional uranium mill	Under general NRC license, in DOE LTSP program	NRC	UMTRCA Title II	2600000	Dry Tonnes	17
Wyoming	Bear Creek	Bear Creek Uranium Co	Conventional uranium mill	Not yet on LTSP. DP approved 5/1989; reclamation completion by 2010	NRC	UMTRCA Title II	4300000	Dry Tonnes	N/A

[244] WA – State of Washington

Annex D-3A   Uranium Mill Tailings and Related Sites

State	Site Name/Location	Licensee	Type	Status	Regulator	Regulatory Program	Quantity of Contaminated Material	Quantity Units	Total ^{226}Ra-Activity (TBq)
	Gas Hills	American Nuclear Corporation	Conventional uranium mill	Not yet on LTSP. DP approved 10/1998; reclamation completion by 2011	NRC	UMTRCA Title II	7300000	Dry Tonnes	N/A
	Gas Hills	Pathfinder Mines Corp – Lucky MC	Conventional uranium mill	Not yet on LTSP. Revised DP approved 6/1996, revised in 1998.	NRC	UMTRCA Title II	10600000	Dry Tonnes	N/A
	East Gas Hills	Umetco Minerals Corp	Conventional uranium mill	Not yet on LTSP. Revised DP approved 4/2001(soil)	NRC	UMTRCA Title II	7300000	Dry Tonnes	N/A
	Highlands	Exxon Mobil Corp	Conventional uranium mill	Not yet on LTSP. DP approved 1990; reclamation completion in 2010	NRC	UMTRCA Title II	10300000	Dry Tonnes	N/A
	Christensen Ranch	Uranium One	In situ site	Application under review to restart facility	NRC	UMTRCA Title II	N/A		N/A
	Moore Ranch	Uranium One	In situ site	Licensed; Under Construction	NRC	UMTRCA Title II	N/A		N/A
	Hank & Nichols	Uranerz Energy	In situ site	Licensed; Under Construction	NRC	UMTRCA Title II	N/A		N/A
	Lost Creek	Lost Creek	In situ site	Licensed; Under Construction	NRC	UMTRCA Title II	N/A		N/A
	Shirley Basin	Pathfinder Mines Corp	Conventional uranium mill	Not yet on LTSP. Revised DP approved 12/1997; candidate for restart	NRC	UMTRCA Title II	7400000	Dry Tonnes	N/A
	Shirley Basin – South	DOE	Conventional uranium mill	Under general NRC license, in DOE LTSP program	NRC	UMTRCA Title II	6300000	Dry Tonnes	N/A
	Smith Ranch – Highland	Power Resources, Inc.	In situ site	Operating	NRC	UMTRCA Title II	N/A		N/A
	Split Rock	Western Nuclear Inc.	Conventional uranium mill	Not yet on LTSP. DP approved 1997	NRC	UMTRCA Title II	7000000	Dry Tonnes	N/A

Annex D-3A   Uranium Mill Tailings and Related Sites

State	Site Name/Location	Licensee	Type	Status	Regulator	Regulatory Program	Quantity of Contaminated Material	Quantity Units	Total ^{226}Ra-Activity (TBq)
	Spook	DOE	Conventional mill and surface residual radioactive material disposal cell	Under general NRC license, in DOE LTSP program	NRC	UMTRCA Title I	1500000	m^3	N/A
	Sweetwater	Kennecott Uranium Co	Conventional uranium mill	On standby status	NRC	UMTRCA Title II	2100000	Dry Tonnes	N/A

Sources: http://www.lm.doe.gov/pro_doc/references/fframework.htm;   http://www.radiationcontrol.utah.gov/
http://www.eia.doe.gov/cneaf/nuclear/dupr.html;   http://www.doh.wa.gov/ehp/rp/waste/dmchm.htm

U.S. Fourth National Report-Joint Convention on the Safety of Spent Fuel Management and on the Safety of Radioactive Waste Management

Annex D-3B Expected New Uranium Recovery Facility Applications / Restarts / Expansions[245]

Company	Site	Design type	Estimated Application Date	State	Letter of Intent
Fiscal 2007 Applications					
Uranium One	Willow Creek (formerly Christensen Ranch)	ISR - Restart	Received April 2007	WY	None
Cameco (Crow Butte Resources, Inc.)	North Trend	ISR - Expansion	Received June 2007	NE	None
Cameco (Crow Butte Resources, Inc.)	Plant Upgrade	ISR - Expansion	Received October 2006	NE	None
Fiscal 2008 Applications					
Lost Creek ISR, LLC	Lost Creek	ISR - New	Resubmitted March 2008	WY	05/23/07
Uranerz Energy Corporation	Hank and Nichols	ISR - New	Received December 2007	WY	06/27/07
Uranium One	Moore Ranch	ISR - New	Received October 2007	WY	05/31/07
Uranium One	Jab and Antelope	ISR - New	Received July 2008	WY	05/31/07
Fiscal 2009 Applications					
Powertech Uranium Corporation	Dewey Burdock	ISR - New	Resubmitted August 2009	SD	01/26/07
Fiscal 2010 Applications					
Cameco (Crow Butte Resources, Inc.)	Three Crow	ISR - Expansion	Received August 2010	NE	01/11/10
Fiscal 2011 Applications					
Strata Energy, Inc.	Ross	ISR - New	December 2010	WY	01/08/10
Uranium One	Allemand-Ross	ISR - Expansion	May 2011	WY	10/08/10
Uranium Energy Corporation	Grants Ridge	Heap Leach - New	June 2011	NM	01/15/10
Lost Creek ISR, LLC	Lost Creek	ISR - Expansion	June 2011	WY	01/06/10
Cameco (Power Resources, Inc.)	Smith Ranch/Highland CPP	ISR - Expansion	July 2011	WY	01/1410
Titan Uranium USA, Inc.	Sheep Mountain	Heap Leach - New	July 2011	WY	11/11/10
Cameco (Crow Butte Resources, Inc.)	Marsland	ISR - Expansion	September 2011	NE	11/09/10
Uranium One	Ludeman	ISR – New	Withdrawn, Resubmit 09/11	WY	10/08/10
Wildhorse Energy	West Alkali Creek	ISR - New	TBD	WY	01/07/10

245 Note that some of these sites are also listed in Annex D-3A

Annex D-3B Expected New Uranium Recovery Facility Applications / Restarts / Expansions[245]

Company	Site	Design type	Estimated Application Date	State	Letter of Intent
Fiscal 2012 Applications					
AUC LLC	Reno Creek	ISR - New	December 2011	WY	11/03/10
UR-Energy Corp.	Lost Soldier	ISR - Expansion	March 2012	WY	11/01/10
Neutron Energy	Juan Tafoya	Conv. - New	June 2012	NM	11/16/10
The Bootheel Project LLC	Bootheel	ISE – New (Satellite)	July 2012	WY	08/09/10
Strathmore Minerals Corporation	Gas Hills	Conv. - New	September 2012	WY	11/19/10
Strathmore Minerals Corporation	Roca Honda	Conv. - New	September 2012	NM	11/19/10
Uranium Company of Nevada, LLC	Apex Hill	Conv. - New	September 2012	NV	11/11/10
Rio Grande Resources	Mt. Taylor	Conv. - New	TBD	NM	11/10/10
Fiscal 2013 Applications					
Cameco (Power Resources, Inc.)	Ruby Ranch	ISR - Expansion	FY 2013	WY	01/14/10
Other Major Licensing Actions					
Cameco (Crow Butte Resources, Inc.)	Crawford, NE	ISR – License Renewal	Received December 2007	NE	
Uranium One	Irigaray/Christensen Ranch	ISR – License Renewal	Received May 2008	WY	
Cameco (Power Resources, Inc.)	Smith Ranch/Highland	ISR – License Renewal	August 2010	WY	
Hydro Resources, Inc.	Crownpoint	ISR – License Renewal	Received August 2002, on hold until 2011	NM	
Cameco (Power Resources, Inc.	North Butte	ISR – Ops Plan	March 2011	WY	09/09/10
Cameco (Power Resources, Inc.	Gas Hills	ISR – Ops Plan	June 2011	WY	09/09/10
Cameco (Power Resources, Inc.	Ruth	ISR – Ops Plan	June 2013	WY	09/09/10

Source: www.nrc.gov/info-finder/materials/uranium/ur-projects-list-public.pdf (last updated 11/18/2010)

ISR = In situ Recovery
Conv. = Conventional Mill

Annex D-4 Pending and Ongoing DOE Decommissioning and Remediation Projects[246]

State	Site	Historic Mission	Number of Nuclear/Radioactive Facilities Decommissioning	Number of Release Sites (Remediation)
California	Energy Technology Engineering Center	Research, Development & Testing	2	10
	Stanford Linear Accelerator Center	Research, Development & Testing		34
Idaho	Idaho Site	Defense, Research, Development & Testing	97	116
Illinois	Argonne National Laboratory	Research, Development & Testing	2	
Kentucky	Paducah Gaseous Diffusion Plant	Enrichment	39	123
Nevada	Nevada National Security Site and off-site test locations	Defense (Weapons Testing)		976
New Mexico	Los Alamos National Laboratory	Defense, Research, Development & Testing	105	663
	Sandia National Laboratories-New Mexico	Defense, Research, Development & Testing		1
New York	Brookhaven National Laboratory	Research, Development & Testing	2	
	Separations Process Research Unit	Research, Development & Testing	4	6
	West Valley Demonstration Project	Commercial Reprocessing	20	
Ohio	Portsmouth Gaseous Diffusion Plant	Enrichment	33	1
South Carolina	Savannah River Site	Defense	212	146
Tennessee	Oak Ridge Reservation	Defense, Research, Development, & Testing	67	285
Washington	Hanford Site	Defense	480	1496
Total			**1063**	**3857**

[246] Source: *Office of Environmental Management FY 2011 Congressional Budget*. Reflects remaining decommissioning and remediation projects as of September 30, 2009.

U.S. Fourth National Report-Joint Convention on the Safety of Spent Fuel Management and on the Safety of Radioactive Waste Management

Annex D-5 Formerly Utilized Sites Remedial Action Program Sites in Progress

State	Site	Status
Connecticut	Combustion Engineering Site (Windsor)	Ongoing Remediation
Indiana	Joslyn Manufacturing and Supply Company (Fort Wayne)	Currently Under Site Investigation
Iowa	Iowa Army Ammunition Plant (Middletown)	Ongoing Remediation
Maryland	W.R. Grace Site (Baltimore)	Remedial Investigation
Massachusetts	Shpack Landfill (Norton/Attleboro)	Ongoing Remediation
Missouri	Latty Avenue Properties (Hazelwood)	Ongoing Remediation
	St. Louis Airport Site (St. Louis)	Remediation Complete
	St. Louis Airport Site Vicinity Properties (St. Louis)	Ongoing Remediation
	St. Louis Downtown Site (St. Louis)	Ongoing Remediation
New Jersey	Maywood Chemical Superfund Site (Maywood)	Ongoing Remediation
	Wayne Interim Storage Site (Wayne)	Remediation Complete
	Middlesex Sampling Plant (Middlesex)	Ongoing Remediation
	DuPont Chamber Works (Deepwater)	Ongoing Remediation
New York	Niagara Falls Storage Site (Lewiston)	Ongoing Remediation
	Former Linde Air Products (Tonawanda)	Ongoing Remediation
	Guterl Specialty Steel (Lockport)	Remedial Investigation
	Seaway Industrial Park (Tonawanda)	Ongoing Remediation
	Colonie Site (Colonie)	Ongoing Remediation
	Sylvania Corning Plant (Hicksville)	Remedial Investigation
Ohio	Luckey Site (Luckey)	Ongoing Remediation
	Painesville Site (Painesville)	Ongoing Remediation
	Harshaw Chemical Company (Cleveland)	Remedial Investigation
Pennsylvania	Shallow Land Disposal Area (Parks Township)	Ongoing Remediation
	Superior Steel (Canegie)	Remedial Investigation to be scheduled

Source: US Army Corps of Engineers, *Formerly Utilized Sites Remedial Action Program Update*, October 2010
Some of these sites are also included in the Materials Decommissioning Program (Annex D-6)

Annex D-6 Decommissioning of Complex Licensed Materials Sites[247]

State	Installation	Location	Decommissioning Status[248]
NRC Regulated Sites			
California	Hunter's Point Naval Shipyard[249]	San Francisco	Estimated closure to be determined
	McClellan (former Air Force Base)3	Sacramento	Estimated closure to be determined
Connecticut	ABB Prospects	Windsor	Estimated closure in 12/2011, under unrestricted release
	UNC Naval Products (a.k.a. United Nuclear)	New Haven	Estimated closure in 9/2011, under unrestricted release
Hawaii	Pohakuloa Training Area (Army)	Hilo/Kawaihae	Estimated closure to be determined
	Schofield Army Barracks	Oahu	Estimated closure to be determined
Indiana	Jefferson Proving Ground (Department of the Army)	Madison	Estimate closure after 12/2013 under restricted release.
Maryland	Beltsville Agricultural Research Laboratory	Beltsville	Estimated closure in 2011, under unrestricted release
Michigan	AAR Manufacturing Group, Inc.	Livonia	Estimated closure in 9/2012, under restricted release
	NWI Breckenridge	Breckenridge	Estimated closure in 2011, under unrestricted release
Missouri	ABC Labs	Columbia	Estimated closure in 9/2011, under unrestricted release
	Mallinckrodt Chemical Inc.	St. Louis	Estimated closure in 1/2012, under unrestricted release
	Sigma-Aldrich	Maryland Heights	Estimated closure in 6/2011, under unrestricted release
	Westinghouse Electric Corp. (Hematite Facility)	Jefferson City	Estimated closure in 12/2013, under unrestricted release
New Jersey	Stepan Chemical Company[250]	Maywood	Estimated closure in 9/2014, under unrestricted release
	Shieldalloy Metallurgical Corp.[251]	Newfield	Court of Appeals vacated site transfer – Pending/TBD
New York	West Valley Demonstration Project	West Valley	Estimated closure to be determined
Oklahoma	FMRI (Fansteel), Inc.	Muskogee	Estimate closure after 6/2023, under unrestricted release
	Kerr-McGee – Cimarron	Cimarron	Estimated closure after 1/2017 under unrestricted release
Pennsylvania	Babcock & Wilcox SLDA[252]	Vandergrift	Estimated closure in 9/2015, under restricted release

---

[247] Source: NUREG 1814.

[248] Unspecified closure dates pending resolution of site-specific regulatory provisions; e.g., financial assurance, waste management arrangements, etc.

[249] The Navy's Hunter's Point Shipyard site and the Air Force's McClellan site are being remediated by the Navy and Air Force, respectively, under the required CERCLA process and EPA oversight. NRC has not licensed these sites, but has approved a "limited approach" and will rely on the ongoing CERCLA process and EPA oversight.

[250] Although New Jersey is now a licensed Agreement State, the Stepan site remained under NRC jurisdiction because of the presence of FUSRAP material.

[251] Transfer of regulatory authority of the Shieldalloy site was set aside because of a judicial appeal.

[252] NRC retains regulatory authority, including decommissioning phase, at sites having special nuclear material in quantities sufficient to form a critical mass.

Annex D-6 Decommissioning of Complex Licensed Materials Sites[253]

State	Installation	Location	Project Completion Status
California	General Atomics	San Diego	Awaiting facility release prior to termination
	Excel Research Services, Inc.	Fresno	Ongoing
	Halaco	Oxnard	EPA Superfund Site – TBD
	The Boeing Company	Simi Valley	Dual jurisdiction and mixed waste have complicated the release
	Providencia Holdings, Inc.	Burbank	Ongoing
	Chevron Mining, Inc. (Formerly Molycorp)	Mountain Pass	Ongoing
	AeroJet Ordnance Co.	Chino	Ongoing
	Isotope Specialties	Burbank	Ongoing
	Magnesium Alloy Products	Compton	TBD
Colorado	Colorado School of Mines Research Institute Table Mtn.	Golden	TBD – delay for financial assurance issues
	Colorado School of Mines Research Institute Creekside	Golden	TBD – delay for financial assurance issues
	Redhill Forest-domestic water treatment	Fairplay	TBD – delay for financial assurance issues.
	Clean Harbors	Deer Trail	TBD – Current legal dispute with local government about site certificate of designation
Florida	Iluka Resources, Inc.	Green Cove Springs	TBD
Illinois	Spectrulite Consortium	Madison	Pending submittal of final certifications and cost recovery fees
	Chicago Magnesium	Blue Island	2011
	TRONOX-Rare Earths & Thorium (Formerly Kerr-McGee)	West Chicago	TBD – pending results from groundwater contamination monitoring and corrective action
Kansas	Air Capitol Dial	Wichita	TBD
	Aircraft Instrument & Development/RC Allen Instruments	Wichita	TBD
	Century Instruments Corporation	Wichita	TBD
	Instrument and Flight Research	Wichita	TBD
	Kelley Instruments, Inc.	Wichita	TBD
	Instrument, Inc.	Wichita	TBD
	Shpack Landfill	Norton	TBD – complicated by multiple jurisdictions
Massachusetts	BASF (Formerly Engelhard Corporation)	Plainville	TBD

253 Source: NUREG 1814.

Annex D-6 Decommissioning of Complex Licensed Materials Sites[253]

Agreement State Regulated Sites

State	Installation	Location	Project Completion Status
	Starmet Corp. (Formerly Nuclear Metals)	Concord	TBD
	Wyman Gordon Co.	North Grafton	TBD
	Texas Instruments	Attleboro	TBD
	Norton/St. Gobain	Worcester	TBD
Nebraska	University of Nebraska Disposal Trenches	Mead	Ongoing; U.S. Army Corp of Engineers performing under EPA oversight
Ohio	Metallurg Vanadium Corp. (Formerly Shieldalloy Metallurgical Corp.)	Cambridge	Ongoing
	Ineos USA, LLC (Formerly BP Chemical)	Lima	12/2020
	Advanced Medical Systems, Inc.	Cleveland	12/2015
Oregon	TDY Industries Dba Wah Chang	Albany	TBD
	PCC Structurals, Inc.	Portland	TBD
	Curtis-Wright	Cheswick	2011
	Karnish Instruments	Lock Haven	TBD
	Molycorp, Inc. (Washington)	Washington	TBD
	Superbolt (formerly Superior Steel)	Carnegie	TBD
	Quehanna (formerly Permagrain Products, Inc.)	Karthaus	TBD
Pennsylvania	Safety Light Corporation	Bloomsburg	TBD - Transferred to EPA - financial assurance remains the key issue
	Strube Incorporated	Lancaster County	TBD
	Westinghouse Electric Corp. (Waltz Mill)	Madison	TBD
	Whittaker Corporation	Greenville	Ongoing

Annex D-7 NRC-Licensed Power and Demonstration Reactors Under Decommissioning

State	Facility	Reactor Type	Power	D&D Status
Commercial Power Reactors				
California	Humboldt Bay 3	Boiling Light-Water Reactor	63 MWe	SAFSTOR
	San Onofre – Unit 1	Pressurized Light-Water Reactor	436 MWe	DECON
	Vallecitos BWR	Boiling Light-Water Reactor	5 MW	SAFSTOR
Connecticut	Millstone – Unit 1	Boiling Light-Water Reactor	660 MWe	SAFSTOR
Illinois	Dresden – Unit 1	Boiling Light-Water Reactor	200 MWe	SAFSTOR
	Zion – Unit 1	Pressurized Light-Water Reactor	1040 MWe	SAFSTOR
	Zion – Unit 2	Pressurized Light-Water Reactor	1040 MWe	SAFSTOR
Michigan	Fermi – Unit 1	Liquid Metal Fast Breeder Reactor	61 MWe	SAFSTOR
New York	Indian Point – Unit 1	Pressurized Light-Water Reactor	257 MWe	SAFSTOR
Pennsylvania	Peach Bottom – Unit 1	High Temperature Gas Reactor	40 MWe	SAFSTOR
	Three Mile Island – Unit 2	Pressurized Light-Water Reactor	792 MWe	Monitored SAFSTOR
Virginia	Nuclear Ship Savannah	Pressurized Light-Water Reactor	80 MW	SAFSTOR
Wisconsin	La Crosse	Boiling Light-Water Reactor	50 MWe	SAFSTOR
Research and Test Reactors				
Arizona	University of Arizona	TRIGA	110 kW	DECON – pending
California	General Atomics	TRIGA Mark F	1500 kW	DECON Approved
	General Atomics	TRIGA Mark I	250 kW	DECON Approved
	General Electric Co.	GETR (Tank)	50 MW	SAFSTOR Possession Only
	General Electric Co.	VESR	2 MW	SAFSTOR Possession Only
Illinois	University of Illinois	TRIGA	1500 kW	DECON Approved
Ohio	National Aeronautics and Space Administration	Tank	60 MW	DECON Approved
	National Aeronautics and Space Administration	Mockup	100 kW	DECON Approved
Massachusetts	Worcester Polytechnic Institute	Pool	10 kW	DECON – pending
Michigan	Ford Nuclear Reactor	Pool	2 MW	DECON Approved
Nebraska	Veterans Administration	TRIGA-Mark I	20 kW	DECON Amendment
New York	University of Buffalo	Pool	2 MW	SAFSTOR Possession Only

Annex E-1 NRC Guidance
NRC provides guidance on acceptable methods for meeting its regulatory requirements. Guidance documents, such as regulatory guides or staff technical positions, are not a substitute for regulations. Compliance with guidance is not required. Methods, analysis, and solutions different from guidance are also acceptable if they demonstrate meeting actual regulatory requirements. Some examples of guidance include:

HLW Management

NUREG-1804, Revision 2, *Yucca Mountain Review Plan (Final Report) July 2003*
Regulatory Guide 3.69, *Topical Guidelines for the Licensing Support System*, Revision 1, June 2004
NUREG-1563, *Branch Technical Position on the Use of Expert Elicitation in the HLW Program,* issued November 1996
Interim Staff Guidance (ISG) Used by the HLW Repository Safety Staff:
• ISG-01, *Review Methodology for Seismically Initiated Event Sequences*
• ISG-02, *Preclosure Safety Analysis — Level of Information and Reliability Estimation*
• ISG-03, *Preclosure Safety Analysis — Dose Performance Objectives and Radiation Protection Program*
• ISG-04, *Preclosure Safety Analysis — Human Reliability Analysis*

LLW Management

Regulatory Guide 4.20, *Constraint on Releases of Airborne Radioactive Materials to The Environment For Licensees Other Than Power Reactors*
Regulatory Guide 4.18, *Standard Format and Content of Environmental Reports for Near-Surface Disposal of Radioactive Waste*, June 1983
NUREG-1200, *Standard Review Plan for the Review of a License Application for a Low-Level Radioactive Waste Disposal Facility*, Revision 3, April 1994
NUREG-1300, *Environmental Standard Review Plan for the Review of a License Application for a Low-Level Radioactive Waste Disposal Facility*
NUREG-1199, *Standard Format and Content of a License Application for a Low-Level Radioactive Waste Disposal Facility*, Revision 2. January 1991
NUREG-1241, *Licensing of Alternative Methods of Disposal of Low-Level Radioactive Waste*
NUREG-1573, *A Performance Assessment Methodology for Low-Level Radioactive Waste Disposal Facilities*
Regulatory Guide 4.19, *Guidance for Selecting Sites for Near-Surface Disposal of Low-Level Radioactive Waste*, August 1988

Uranium Recovery

NUREG-1724, *Standard Review Plan for the Review of DOE Plans for Achieving Regulatory Compliance at Sites with Contaminated Ground Water Under Title I of the Uranium Mill Tailings Radiation Control Act: Draft Report for Comment*, June 2000
NUREG-1623, *Design of Erosion Protection for Long-Term Stabilization*, September 2002.
NUREG-1620, Rev. 1. *Standard Review Plan for the Review of a Reclamation Plan for Mill Tailings Sites Under Title II of the Uranium Mill Tailings Radiation Control Act*, June 2003
NUREG-1569, *Standard Review Plan for In Situ Leach Uranium Extraction License Applications*, June 2003.
Uranium Mill In Situ Leach Uranium Recovery, and 11e.(2) Byproduct Material Disposal Site Decommission Inspection, (Procedure 87654), March 2002
Regulatory Guide 3.5, *Standard Format and Content of License Applications for Uranium Mills*
Regulatory Guide 3.8, *Preparation of Environmental Reports for Uranium Mills*

189

U.S. Fourth National Report-Joint Convention on the Safety of Spent Fuel Management and on the Safety of Radioactive Waste Management

Annex E-1 NRC Guidance
Regulatory Guide 3.11, *Design, Construction, and Inspection of Embankment Retention Systems at Uranium Recovery Facilities*
Regulatory Guide 3.46, *Standard Format and Content of License Applications, Including Environmental Reports, for In Situ Uranium Solution Mining*
Regulatory Guide 3.51, *Calculational Models for Estimating Radiation Doses to Man from Airborne Radioactive Materials Resulting from Uranium Milling Operations*
Regulatory Guide 3.56, *General Guidance for Designing, Testing, Operating, and Maintaining Emission Control Devices at Uranium Mills*
Regulatory Guide 3.59, *Methods for Estimating Radioactive and Toxic Airborne Source Terms for Uranium Milling Operations*
Regulatory Guide 3.63, *Onsite Meteorological Measurement Program for Uranium Recovery Facilities — Data Acquisition and Reporting*
Regulatory Guide 3.64, *Calculation of Radon Flux Attenuation by Earthen Uranium Mill Tailings Covers*
Regulatory Guide 4.14, *Radiological Effluent and Environmental Monitoring at Uranium Mills*
Regulatory Guide 8.11, *Applications of Bioassay for Uranium*
Regulatory Guide 8.22, *Bioassay at Uranium Mills*
Regulatory Guide 8.30, *Health Physics Surveys in Uranium Recovery Facilities*
Regulatory Guide 8.31, *Information Relevant to Ensuring that Occupational Radiation Exposures at Uranium Recovery Facilities Will Be as Low as Is Reasonably Achievable*
NUREG-1748, *Environmental Review Guidance for Licensing Actions Associated with NMSS Programs*
Decommissioning
Primary Decommissioning Guidance Documents
• NUREG-1757, *Consolidated Decommissioning Guidance*, Volumes 1-3
• NUREG-1700, *Standard Review Plan for Evaluating Nuclear Power Reactor License Termination Plans*, April 2003
Regulatory Guide 1.159, *Assuring the Availability of Funds for Decommissioning Nuclear Reactors*, 2003
Regulatory Guide 1.202, *Standard Format and Content of Decommissioning Cost Estimates for Nuclear Power Reactors*, 2005
Regulatory Guide 4.15, *Quality Assurance for Radiological Monitoring Programs (Inception through Normal Operations to License Termination) – Effluent Streams and the Environment*, 2007
Multi-Agency Radiation Survey and Assessment of Materials and Equipment Manual (MARSAME); A Supplement to MARSSIM (NUREG-1574, Supp. 1)
Background as a Residual Radioactivity Criterion for Decommissioning (NUREG-1501)
Solubility and Leaching of Radionuclides in Site Decommissioning Management Plan (SDMP) Soil and Ponded Wastes (NUREG/CR-6821)
Information on Hydrologic Conceptual Models, Parameters, Uncertainty Analysis, and Data Sources for Dose Assessments at Decommissioning Sites (NUREG/CR-6695)
Solubility and Leaching of Radionuclides in Site Decommissioning Management Plan (SDMP) Slags (NUREG/CR-6656)
Solubility and Leaching of Radionuclides in Site Decommissioning Management Plan (SDMP) Slags (NUREG/CR-6632)
Revised Analyses of Decommissioning Reference Non-Fuel-Cycle Facilities (NUREG/CR-6477)
Radiological Surveys for Controlling Release of Solid Materials (NUREG-1761)
Standard Review Plan for Decommissioning Cost Estimates for Nuclear Power Reactors (NUREG-1713)

190

U.S. Fourth National Report-Joint Convention on the Safety of Spent Fuel Management and on the Safety of Radioactive Waste Management

Annex E-1 NRC Guidance
Research and Test Reactor Inspection Program (IMC 2545)
Decommissioning Power Reactor Inspection Program (IMC 2561)
NUREG-1556, *Consolidated Guidance About Nuclear Materials*, Vol 1-20
Regulatory Guide 1.184, *Decommissioning Of Nuclear Power Reactors*, July 2000
Regulatory Guide 1.185, *Standard Format and Content for Post-shutdown Decommissioning Activities*, July 2000
NUREG-1575, *Multi-Agency Radiation Survey and Site Investigation Manual, Revision 1.* August 2001
Regulatory Guide 1.179, *Standard Format and Content of License Termination Plans for Nuclear Power Reactors*, Revision 1, June 2011
NUREG/CR-6477, *Revised Analyses of Decommissioning Reference -Non-Fuel-Cycle Facilities*, December 2002
NUREG-1628, *Staff Responses to Frequently Asked Questions Concerning Decommissioning of Nuclear Power Reactors*, June 2000
NUREG-0586, *Generic Environmental Impact Statement on Decommissioning of Nuclear Facilities* (also NUREG-0586, Supplement 1, Vols. 1 & 2)
NUREG-1496, *Generic Environmental Impact Statement in Support of Rulemaking on Radiological Criteria for License Termination of NRC-Licensed Nuclear Facilities*, Vols. 1-3, U.S. Nuclear Regulatory Commission, Washington, D.C.
Spent Fuel Management
NUREG-1536, *Standard Review Plan for Dry Cask Storage Systems*
NUREG-1567, *Standard Review Plan for Spent Fuel Dry Storage Facilities*
NUREG-1609, *Standard Review Plan for Transportation Packages for Radioactive Material, Supplement 1, Standard Review Plan for Transportation Packages for MOX-Radioactive Material and Supplement 2, Standard Review Plan for Transportation Packages for Irradiated Tritium-Producing Burnable Absorber Rods (TPBARs)*
NUREG-1617, *Standard Review Plan for Transportation Packages for Spent Nuclear Fuel, Supplement 1, Standard Review Plan for Transportation Packages for MOX Spent Nuclear Fuel*
NUREG-1927, *Standard Review Plan for Renewal of Spent Fuel Dry Cask Storage System Licenses and Certificates of Compliance*
Interim Staff Guidance:
SFST-ISG-1, *Damaged Fuel*
SFST-ISG-2, *Fuel Retrievability*
SFST-ISG-3, *Post Accident Recovery and Compliance with 10 CFR 72.122(l)*
SFST-ISG-4, *Cask Closure Weld Inspections*
SFST-ISG-5, *Confinement Evaluation*
SFST-ISG-6, *Establishing Minimum Initial Enrichment for the Bounding Design Basis Fuel Assembly(s)*
SFST-ISG-7, *Potential Generic Issue Concerning Cask Heat Transfer in a Transportation Accident*
SFST-ISG-8, *Burnup Credit in the Criticality Safety Analyses of PWR Spent Fuel in Transport and Storage Casks*
SFST-ISG-9, *Storage of Components Associated with Fuel Assemblies*
SFST-ISG-10, *Alternatives to the ASME Code*
SFST-ISG-11, *Cladding Considerations for the Transportation and Storage of Spent Fuel*
SFST-ISG-12, *Buckling of Irradiated Fuel Under Bottom End Drop Conditions*
SFST-ISG-13, *Real Individual*

191

U.S. Fourth National Report-Joint Convention on the Safety of Spent Fuel Management and on the Safety of Radioactive Waste Management

Annex E-1 NRC Guidance
SFST-ISG-14, *Supplemental Shielding*
SFST-ISG-15, *Materials Evaluation*
SFST-ISG-16, *Emergency Planning*
SFST-ISG-17, *Interim Storage of Greater Than Class C Waste*
SFST-ISG-18, *The Design/Qualification of Final Closure Welds on Austenitic Stainless Steel Canisters as Confinement Boundary for Spent Fuel Storage and Containment Boundary for Spent Fuel Transportation*
SFST-ISG-19, *Moderator Exclusion Under Hypothetical Accident Conditions and Demonstrating Subcriticality of Spent Fuel Under the Requirements of 10 CFR 71.55(e)*
SFST-ISG-20, *Transportation Package Design Changes Authorized Under 10 CFR Part 71 Without Prior NRC Approval*
SFST-ISG-21, *Use of Computational Modeling Software*
SFST-ISG-22, *Potential Rod Splitting Due to Exposure to an Oxidizing Atmosphere During Short-Term Cask Loading Operations in LWR or Other Uranium Oxide Based Fuel*
SFST-ISG-23, *Application of ASTM Standard Practice C1671-07 When Performing Technical Reviews of Spent Fuel Storage and Transportation Packaging Licensing Actions*
SFST-ISG-25, *Pressure Test and Helium Leakage Test of the Confinement Boundary for Spent Fuel Storage Canister*
Regulatory Guide 1.13, *Spent Fuel Storage Facility Design Basis*, Rev. 2, March 2007
Regulatory Guide 3.48, *Standard Format and Content for the Safety Analysis Report for an Independent Spent Fuel Storage Installation or Monitored Retrievable Storage Installation (Dry Storage)*, Rev. 1, August 1989
Regulatory Guide 3.50, *Standard Format and Content for a License Application To Store Spent Fuel and High-Level Radioactive Waste (Draft FP 907-4 published 3/1981)* Rev.1. September 1989
Regulatory Guide 3.53, *Applicability of Existing Regulatory Guides to the Design and Operation of an Independent Spent Fuel Storage Installation*, July 1982
Regulatory Guide 3.54, *Spent Fuel Heat Generation in an Independent Spent Fuel Storage Installation*, Rev. 1. January 1999
Regulatory Guide 3.60, *Design of an Independent Spent Fuel Storage Installation (Dry Storage)*, March1987
Regulatory Guide 3.61, *Standard Format and Content for a Topical Safety Analysis Report for a Spent Fuel Dry Storage Cask*, February 1989
Regulatory Guide 3.62, *Standard Format and Content for the Safety Analysis Report for Onsite Storage of Spent Fuel Storage Casks*, February 1989
Regulatory Guide 3.72, *Guidance for Implementation of 10 CFR 72.48, Changes, Tests, and Experiments*, March 2001
Regulatory Guide 3.73, *Site Evaluations and Design Earthquake Ground Motion for Dry Cask Independent Spent Fuel Storage and Monitored Retrievable Storage Installations*
Regulatory Guide 4.21, *Minimization of Contamination and Radioactive Waste Generation: Life-Cycle Planning*, June 2008
Regulatory Guide 7.4, *Leakage Tests on Packages for Shipment of Radioactive Materials*, June 1975
Regulatory Guide 7.6, *Design Criteria for the Structural Analysis of Shipping Cask Containment Vessels*, Rev. 1, March 1978
Regulatory Guide 7.7, *Packaging and Transportation of Radioactively Contaminated Biological Materials*, June 1974
Regulatory Guide 7.8, *Load Combinations for the Structural Analysis of Shipping Casks for Radioactive Material*, Rev. 1, March 1989

Annex E-1 NRC Guidance
Regulatory Guide 7.9, *Standard Format and Content of Part 71 Applications for Approval of Packages for Radioactive Material*, Rev. 2, March 2005
Regulatory Guide 7.10, *Establishing Quality Assurance Programs for Packaging Used in Transport of Radioactive Material*, Rev. 2, March 2005
Regulatory Guide 7.11, *Fracture Toughness Criteria of Base Material for Ferritic Steel Shipping Cask Containment Vessels with a Maximum Wall Thickness of 4 Inches (0.1 m)*, June 1991
Regulatory Guide 7.12, *Fracture Toughness Criteria of Base Material for Ferritic Steel Shipping Cask Containment Vessels with a Wall Thickness Greater than 4 Inches (0.1 m) But Not Exceeding 12 Inches (0.3 m)*, June 1991

193

U.S. Fourth National Report-Joint Convention on the Safety of Spent Fuel Management and on the Safety of Radioactive Waste Management

Annex F-1 Radiation Protection Guidance	
Federal guidance is a set of guidelines developed by EPA for use by Federal and state agencies responsible for protecting the public from the harmful effects of radiation. Prior to the formation of EPA in 1970, Federal guidance was the responsibility of the Federal Radiation Council. Guidance on radiation protection from EPA comes in two forms:	
Federal Guidance Recommendations, which are signed by the President and usually reflected in Federal regulations for radiation protection of workers or the general public, and	
Federal Guidance Technical Reports, which help standardize radiation dose and risk assessment methodologies.	
Federal Guidance Recommendations	
Radiation Protection Guidance for Federal Agencies, Federal Radiation Council 25 FR 9057 September 26, 1961.	This guidance provides recommendations for population groups exposed to environmental sources of radiation. It provides Radiation Protection Guides; guidance on general principles of control applicable to all environmental radionuclides; and specific guidance in connection with exposure of population groups to ^{226}Ra, ^{131}I, ^{90}Sr, and ^{89}Sr.
Radiation Protection Guidance for Federal Agencies, Federal Radiation Council 25 FR 4402 May 18, 1960.	This guidance provides a general framework for radiation protection and general principles of radiation control based on the annual intake of radioactive materials. These recommendations provide the basis for the control and regulation of radiation exposure during normal peacetime operations. Numerical values for the Radiation Protection Guides, designed to limit the exposure of the whole body and certain organs, are provided.
Radiation Protection Guidance to Federal Agencies for Occupational Exposure, U.S. Environmental Protection Agency 52 FR 2822 January 27, 1987.	This guidance provides general principles, and specifies the numerical primary guides for limiting worker exposure. It applies to all workers who are exposed to radiation in the course of their work, either as employees of institutions and companies subject to Federal regulation or as Federal employees.
Radiation Protection Guidance to Federal Agencies for Diagnostic X-rays, U.S. Environmental Protection Agency 43 FR 4377 February 1, 1978.	This guidance provides recommendations to reduce radiation exposure from the use of diagnostic x-rays. These recommendations, transmitted to the President jointly by EPA and the Department of Health, Education and Welfare were based on two guiding principles: avoidance of unnecessary prescription of x-rays, and use of good technique to minimize radiation exposure.
Underground Mining of Uranium Ore, Federal Radiation Council 34 FR 576 January 15, 1969 35 FR 245 December 18, 1970.	This guidance sets forth recommendations for radiation protection activities as they apply to the underground mining of uranium ore. EPA subsequently reviewed these recommendations and concluded no modification was necessary.
Federal Guidance Technical Reports	
Technical reports summarize current scientific and technical information for radiation dose and risk assessments. Examples of technical reports are:	
Background Material for the Development of Radiation Protection Standards, FGR 5, Federal Radiation Council, July 1964.	This guidance provides background material used in the development of guidance for Federal agencies for (1) planning protective actions to reduce potential doses to the population from radioactive fission products which may contaminate food, and (2) doses at which implementation of protective actions may be appropriate.

194

U.S. Fourth National Report-Joint Convention on the Safety of Spent Fuel Management and on the Safety of Radioactive Waste Management

Annex F-1 Radiation Protection Guidance	
Limiting Values of Radionuclide Intake and Air Concentration and dose Conversion Factor for Inhalation, Submersion, and Ingestion, FGR 11, EPA-520/1-88-020, September 1988.	This guidance provides derived guides (limiting values) of radionuclide intake and air concentration for control of occupational exposure that are consistent with 1987 Federal Guidance. The derived guides serve as the basis for regulations setting upper bounds on the inhalation and ingestion of, and submersion in, radioactive materials in the workplace. The report also includes tables of exposure-to-dose conversion factors for general use in assessing average individual committed doses.
External Exposure to Radionuclides in Air, Water, and Soil, FGR 12, EPA 402-R-93-081, September 1993.	This guidance provides dose coefficients for use by Federal agencies having regulatory responsibilities for protection of members of the public and/or workers.
Cancer Risk Coefficients for Environmental Exposure to Radionuclides, FGR 13, EPA 402-R-99-001, September 1999.	This guidance provides methods and data for estimating risks due to both internal and external radionuclide exposures. The information presented in this report is for use in assessing risks from radionuclide exposure in a variety of applications ranging from environmental impact analyses of specific sites to the general analyses supporting rulemaking.

195

U.S. Fourth National Report-Joint Convention on the Safety of Spent Fuel Management and on the Safety of Radioactive Waste Management

Annex F-2 NRC Participation in Emergency Exercises During 2008-2010		
Date	**Facility/Activity/ Exercise Title**	**Participants**
May 6-9, 2008	Forward Challenge	Multi Agency
June 1-4, 2009	Empire 09	Multi Agency
June 07, 2009	Eagle Horizon COOP	Multi Agency
July 27, 2009	National Level Exercise	Multi Agency
October 27, 2009	Global Nuclear	NRC HQ, Region II
April 26-29, 2010	Liberty RadEx	Multi Agency
May 17-21, 2010	National Level Exercise	Multi Agency
September 15, 2010	BWNOG	NRC HQ, Region II

Annex F-3 Requirements for Notifying NRC of Emergency and Non-Emergency Events[254]

Specific requirements for NRC-licensed radioactive materials[255]

Standards for Protection Against Radiation (10 CFR Part 20):

- Section 20.1906 Procedures for receiving and opening packages. [Section 20.1906(d) notification of removable radioactive surface contamination]
- Section 20.2201 Reports of theft or loss of licensed material.
- Section 20.2202 Notification of incidents.
- Section 20.2203 Reports of exposures, radiation levels, and concentrations of radioactive material exceeding the constraints or limits.
- Section 20.2204 Reports of planned special exposures.

Domestic Licensing of Byproduct Material (10 CFR Part 30):

- Section 30.50 Reporting requirements.
- Section 30.55 Tritium reports.

Domestic Licensing of Source Material (10 CFR Part 40):

- Section 40.60 Reporting requirements.
- Section 40.64 Reports. [Section 40.64(c)(2) notification for any attempt of theft or unlawful diversion]

Domestic Licensing of Production and Utilization Facilities (10 CFR Part 50):

- Section 50.72 Immediate notification requirements for operating nuclear power reactors.
- Section 50.73 Licensee event reporting system.

Disposal of High-Level Radioactive Wastes In Geologic Repositories (10 CFR 60.78, Material Control and Accounting Records and Reports).

Disposal of High-Level Radioactive Wastes In A Geologic Repository at Yucca Mountain, Nevada (10 CFR 63.73, Reports of Deficiencies).

Domestic Licensing of Special Nuclear Material (10 CFR Part 70):

- Section 70.50 Reporting requirements.
- Section 70.52 Reports of accidental criticality or loss or theft or attempted theft of special nuclear material.
- Section 70.74 Additional reporting requirements. (Appendix A -Reportable Safety Events)

Appendix A to Part 70--Reportable Safety Events.

Packaging and Transportation of Radioactive Material. (10 CFR 71.95 Reports)

Licensing Requirements for Independent Storage of Spent Fuel and High-Level Radioactive Waste (10 CFR Part 72):

- Section 72.32 Emergency Plan. (8) Notification and coordination.
- Section 72.74 Reports of accidental criticality or loss of special nuclear material.
- Section 72.75 Reporting requirements for specific events and condition.

Physical Protection of Plants and Materials (10 CFR part 73):

- Section 73.27 Notification requirements.
- Section 73.67 Licensee fixed site and in-transit requirements for the physical protection of special nuclear material of moderate and low strategic significance.

---

[254] For more information on NRC Incident Investigation Program, see NRC Management Directive 8.3, NRC Incident Investigation Program accessible at: http://pbadupws.nrc.gov/docs/ML0312/ML031250592.pdf

[255] There are equivalent requirements for the relevant Agreement States.

197

U.S. Fourth National Report-Joint Convention on the Safety of Spent Fuel Management and on the Safety of Radioactive Waste Management

Annex F-3 Requirements for Notifying NRC of Emergency and Non-Emergency Events[254]
• Section 73.71 Reporting of safeguards events.
Appendix G to Part 73--Reportable Safeguards Events
Material Control and Accounting of Special Nuclear Material (10 CFR 74.11 Reports of loss or theft or attempted theft or unauthorized production of special nuclear material).
Certification of Gaseous Diffusion Plants (10 CFR 76.120 Reporting requirements.
Export and Import of Nuclear Equipment and Material (10 CFR 110.50 Terms)
Examples of non-reactor incident reports
NUREG-1405, *Inadvertent Shipment of a Radiographic Source from Korea to Amersham Corporation, Burlington, Massachusetts* (Publication Date: May 1990).
NUREG-1450, *Potential Criticality Accident at the General Electric Nuclear Fuel and Component Manufacturing Facility*, May 29, 1991 (Publication Date: August 1991).
NUREG-1480, *Loss of an Iridium-192 Source and Therapy Misadministration at Indiana Regional Cancer Center Indiana, Pennsylvania on November 16, 1992* (Publication Date: February 1993).
NUREG-1535, *Ingestion of Phosphorus-32 at Massachusetts Institute of Technology, Cambridge, Massachusetts*, Identified on August 19, 1995 (Publication Date: December 1995).
Links to Additional Information on Response to Incidents
Federal Emergency Management Agency (FEMA) State Offices and Agencies of Emergency Management: http://www.fema.gov/about/contact/statedr.shtm
FEMA: http://www.fema.gov/
Department of Energy (DOE): http://energy.gov
U.S. Environmental Protection Agency (EPA): http://www.epa.gov/
Department of Agriculture (USDA): http://www.usda.gov/
Department of Health and Human Services (HHS): http://www.hhs.gov/
Department of State (DOS): http://www.state.gov/
Federal Bureau of Investigation (FBI): http://www.fbi.gov/

Annex F-4. Emergency Preparedness and Planning at Diverse Waste Management Facilities

Geological Repository for Spent Fuel and HLW

NRC requires DOE to develop and be prepared to implement a plan to cope with radiological accidents occurring at the geologic repository operations area, at any time before permanent closure and decontamination or decontamination and dismantlement of surface facilities (10 CFR 60.21(a)(9); 10 CFR 60.131(e) and 10 CFR 63.161). The emergency plan for storage of spent fuel must be based on the criteria of 10 CFR 72.32(b). These criteria require an Emergency Plan including:

- Facility description;
- Types of accidents;
- Classification of accidents;
- Detection of accidents;
- Mitigation of consequences;
- Assessment of releases;
- Responsibilities;
- Notification and coordination;
- Information to be communicated;
- Training;
- Safe condition;
- Exercises;
- Hazardous chemicals;
- Comments on Plan;
- Off-site assistance; and
- Arrangements made for providing information to the public.

LLW Facilities

An applicant must provide a description of the radiation safety program for control and monitoring of radioactive effluents as part of the radiation safety program required for a specific license to dispose of LLW. The objective is to ensure compliance with performance requirements in the regulation (10 CFR Part 61, Subpart C), occupational radiation exposure to comply with 10 CFR Part 20 and control contamination of personnel, vehicles, equipment, buildings, and the disposal site. Both routine operations and accidents are addressed. The program description includes procedures, instrumentation, facilities, and equipment. The regulations specific to emergency planning are in 10 CFR 61.12(k) and §61.13.

The applicant for a near surface disposal site for LLW must propose an emergency response plan. NRC or Agreement States will review this plan to determine whether the licensee would be able to respond to all credible radiological accidents and emergencies consistent with the proposed method of operations. The criteria to assess such a demonstration are in NUREG-1200, *Standard Review Plan for the Review of a license application for a LLW Disposal Facility*. These criteria include:

1) Compliance with 44 CFR Part 350, *Review and Approval of State and Local Radiological Emergency Plans and Preparedness*;
2) Establishing plans to respond to all credible radiological accidents and emergencies consistent with the proposed method of operations; and

199

3) Demonstrating the maximum off-site releases for the most credible accident consistent with the projected source term will yield an off-site dose equivalent of less than 0.1 mSv (0.01 rem) to the whole body and 0.5 mSv (0.05 rem) to the lungs.

The applicant must develop emergency procedures, including interaction with local and State authorities, as well as notification of affected populations where the maximum potential off-site releases yield greater dose equivalents. Such procedures must be developed with knowledge, participation and cooperation of these authorities and affected populations.

The applicant presents this analysis in the safety analysis report provided with the license application; the Standard Format and Content of a License Application for a Low-Level Radioactive Waste Disposal Facility (NUREG-1199, Revision 2) and the Environmental Standard Review Plan for the Review of a License Application for a Low-Level Radioactive Waste Disposal Facility (NUREG-1300) provide guidance for prospective applicants. The accident scenarios addressed include:

- Waste spillage;
- Fire and/or chemical reactions;
- Transportation accidents;
- Nuclear criticality; and
- On-site effects of off-site accidents.

Uranium Recovery Waste Management Facilities

Accidental releases and emergency preparedness are addressed as part of the operational phase of uranium recovery. The perpetual disposal design is required to be robust and not need active maintenance to assure isolation and stability from 200 to 1000 years. Operational considerations for emergency planning during the operational phase are addressed in 10 CFR 40.31(j)(3).

Credible incidents at a uranium milling facility and at a uranium mine would result in minor exposures. The analysis documented in NUREG-1140 estimates a 1-mSv (0.1-rem) effective dose equivalent under the most adverse weather conditions for a fire at a uranium mill. Fires and uranium mill tailings releases (dam failures, pipeline ruptures, etc...) from the late 1950s through the early 1980s are documented in NUREG-1140.

Decommissioning

NRC does not identify a critical radiological accident for decommissioning. Licensees are required to analyze their particular facility and determine the appropriate health and safety measures necessary to maintain worker and public doses within NRC limits. The health and safety plan is provided to NRC as part of the decommissioning or license termination plan (DP or LTP). NRC reviews the plan as part of its review and approval of the DP or LTP.

Role of Inspection and Emergency Preparedness for Decommissioning

Existing Inspection Manual Chapters, Inspection Procedures and Temporary Instructions are applicable and recommended for use in inspection of sites undergoing decommissioning. IP 8850 can be used for emergency preparedness inspection at facilities undergoing decommissioning.

The objectives of this procedure are to ensure the licensee or certificate holder:

- Is complying with regulations and license/certificate requirements related to the processing, control, release, and reporting of information to NRC of radioactive liquid and airborne effluents;
- Is implementing a program to ensure that releases of radioactivity to the environment provide minimal impact on the environment and the public; and
- Maintains adequate management controls for the radiological effluent control and environmental program.

Additional implementation guidance is further detailed in the IP.[256]

---

[256] This document can be downloaded from: http://www.nrc.gov/reading-rm/doc-collections/insp-manual/changenotices/2006/ip88050.doc

LIST OF ACRONYMS AND ABBREVIATIONS

Acronym	Name
ACRS	Advisory Committee on Reactor Safeguards
ADAMS	Agency-wide Documents Access and Management System (NRC)
AEA	Atomic Energy Act of 1954
AEA Section 11e.(2)	Byproduct Material
AEC	Atomic Energy Commission
ALARA	As Low as Reasonably Achievable
ANSI	American National Standards Institute
ANS-8	American Nuclear Society Standards Subcommittee 8
ARARs	Applicable or Relevant and Appropriate Requirements
ASERs	Annual Site Environmental Reports
ASLB	Atomic Safety and Licensing Board
CERCLA	Comprehensive Environmental Response, Compensation, and Liability Act
CFR	Code of Federal Regulations
CoC	Certificate of Compliance
CRCPD	Conference of Radiation Control Program Directors
D&D	Decontamination & Decommissioning
DECON	Immediate Dismantlement
DHS	Department of Homeland Security
DNDO	Domestic Nuclear Detection Office (DHS)
DNFSB	U.S. Defense Nuclear Facilities Safety Board
DOE	U.S. Department of Energy
DOE-HSS	DOE Office of Health, Safety and Security
DOL	U.S. Department of Labor
DoS	U.S. Department of State
DP	Decommissioning Plan
DU	Depleted Uranium
EA	Environmental Assessment
EIS	Environmental Impact Statement
EnPA	Energy Policy Act of 1992
ENTOMB	Entombment
EPA	U.S. Environmental Protection Agency
EPAct05	Energy Policy Act of 2005
EPCRA	Emergency Planning and Community Right-to-Know Act
ERDA	Energy Research and Development Administration
FEMA	Federal Emergency Management Agency
FFA	Federal Facility Agreements

202

U.S. Fourth National Report-Joint Convention on the Safety of Spent Fuel Management and on the Safety of Radioactive Waste Management

Acronym	Name
FR	Federal Register
FTE	Full Time Equivalent
FUSRAP	Formerly Utilized Sites Remedial Action Program
GTCC	Greater-than-Class C Low-Level Waste
GTRI	Global Threat Reduction Initiative
HEU	Highly-Enriched Uranium
HEW	Department of Health, Education and Welfare
HLW	High-Level Waste
HSS	Office of Health, Safety and Security
HWFP	Hazardous Waste Facility Permit
IAEA	International Atomic Energy Agency
ICRP	International Commission on Radiation Protection
IMPEP	Integrated Materials Performance Evaluation Program
INES	International Nuclear Event Scale
INL	Idaho National Laboratory
ISFSI	Independent Spent Fuel Storage Installation
ISR	In Situ Recovery
ITDB	Illicit Trafficking Database
LAW	Low-Activity Waste
LEU	Low-Enriched Uranium
LLD	Lower Limit of Detection
LLEA	Local Law Enforcement Agency
LLRWPA	Low-Level Radioactive Waste Policy Act of 1980
LLRWPAA	Low-Level Radioactive Waste Policy Amendments Act of 1985
LLW	Low-Level Waste
LTP	License Termination Plan
LTSP	Long-Term Surveillance Plan
LVS	License Verifications System
m^3	Cubic Meters
MOU	Memorandum of Understanding
MOX	Mixed Oxide
MRB	Management Review Board
MT	Metric Tons
MTU	Metric Tons Uranium
MTHM	Metric Tons Heavy Metal
NAS	National Academy of Sciences
N/A	Not Applicable or Not Available
NCRP	National Council on Radiation Protection And Measurements
NE	Office of Nuclear Energy

Acronym	Name
NEA	Nuclear Energy Agency (Organization for Economic Co-operation and Development)
NEPA	National Environmental Policy Act
NERI	Nuclear Energy Research Initiative
NESHAPs	National Emission Standards for Hazardous Air Pollutants
NEUP	Nuclear Energy University Program
NNSA	National Nuclear Security Administration
NORM	Naturally Occurring Radioactive Materials
NOV	Notice of Violation
NPL	National Priorities List for CERCLA
NPP	Nuclear Power Plant
NRC	U.S. Nuclear Regulatory Commission
NSTS	National Source Tracking System
NUREG	Nuclear Regulatory Commission Regulation designates a publication prepared by NRC staff
NWPA	Nuclear Waste Policy Act of 1982
NWPAA	Nuclear Waste Policy Amendments Act of 1987
NWTRB	U.S. Nuclear Waste Technical Review Board
OCRWM	Office of Civilian Radioactive Waste Management (DOE)
ORNL	Oak Ridge National Laboratory
OSRP	Off-Site Source Recovery Project
PFS	Private Fuel Storage, LLC
QA	Quality Assurance
RCRA	Resource Conservation and Recovery Act of 1976
R&D	Research and Development
REIRS	Radiation Exposure Information and Reporting System
RH	Remote-handled
ROD	Record of Decision
SAFSTOR	A nuclear facility is safely stored until subsequently decontaminated to permit release for unrestricted use.
SCATR	Source Collection and Threat Reduction Program
SDMP	Site Decommissioning Management Plan
SFP	Spent Fuel Pool
SGI	Safeguards Information
SIP	State Implementation Plans
SSAB	Site-Specific Advisory Boards
Sv	Sievert
Task Force	Task Force on Radiation Source Protection and Security
TEDEs	Total Effective Dose Equivalents
TENORM	Technologically Enhanced NORM

204

U.S. Fourth National Report-Joint Convention on the Safety of Spent Fuel Management and on the Safety of Radioactive Waste Management

Acronym	Name
TRU	Transuranic
UFD	Used Nuclear Fuel Disposition
UMTRCA	Uranium Mill Tailings Radiation Control Act
URL	Underground Research Laboratories
U.S.	United States of America
USACE	U.S. Army Corps of Engineers
U&Th	Uranium and Thorium
WBL	Web-based Licensing
WCS	Waste Control Specialists
WIPP	Waste Isolation Pilot Plant
WIPP LWA	Waste Isolation Pilot Plant Land Withdrawal Act of 1992
WSB	Waste Solidification Building

205

U.S. Fourth National Report-Joint Convention on the Safety of Spent Fuel Management and on the Safety of Radioactive Waste Management

ADDITIONAL REFERENCES

Numerous references to laws, regulations, regulatory guides, standards, and DOE orders are provided throughout this report and are not repeated here (see Table E-1, Table E-2, Annex E-1 and Annex F-1) for brevity. Internet web sites are also provided in Table A-2. The following additional resources were used:

International Atomic Energy Agency, Classification of Radioactive Waste; General Safety Guide Series No. GSG-1, December 28, 2009. Vienna, Austria.

International Atomic Energy Agency, Governmental, Legal and Regulatory Framework for Safety General Safety Requirements Part 1; Series No. GSR Part 1, October 04, 2010. Vienna, Austria.

International Atomic Energy Agency, Guidelines Regarding the Form and Structure of National Reports: Joint Convention on the Safety of Spent Fuel Management and on the Safety of Radioactive Waste Management, INFCIRC/604/Rev.1, 19 July 2006, Vienna, Austria.

International Atomic Energy Agency, Joint Convention on the Safety of Spent Fuel Management and on the Safety of Radioactive Waste Management, INFCIRC/546, Vienna, Austria, December 24, 1997.

U.S. Nuclear Regulatory Commission, NUREG/BR-0216, Radioactive Waste: Production, Storage, Disposal, and Revision 2. Washington DC, USA, May, 2002.

U.S. Nuclear Regulatory Commission, NUREG-1650, The United States of America Fifth National Report for the Convention on Nuclear Safety, Revision 3, Washington DC, USA, September 2010. See: http://www.nrc.gov/reading-rm/doc-collections/nuregs/staff/sr1650/r3/sr1650r3.pdf

U.S. Nuclear Regulatory Commission, Information Digest 2011-2012 Edition (NUREG 1350, Vol. 23). Washington, DC, USA, 2002. August 2011.

U.S. Department of Energy, Energy Information Administration, Report No. DOE/EIA-0592, Decommissioning of U.S. Uranium Production Facilities, Washington, DC, USA, February 1995.

U.S. Department of Energy Waste Management Data, database maintained by Florida International University (users have to register) http://emwims.org/

U.S. Department of Energy, Low-Level Waste Disposal Capacity Report, (2000). http://www.em.doe.gov/stakepages/wmdi_llwtoc.aspx?PAGEID=WMDI

U.S. Department of Energy, DOE/EIS-0375-D, Disposal of Greater-Than-Class C (GTCC) Low-Level Radioactive Waste and GTCC-Like Waste, Volumes 1 and 2, Washington DC, February 2011.

U.S. Department of Energy National Spent Fuel Database (2008). https://inlportal.inl.gov/portal/server.pt/community/national_spent_nuclear_fuel/389/national_spent_nuclear_fuel_-_snf_data

U.S. Department of Energy Grand Junction Office,
http://www.em.doe.gov/bemr/BEMRSites/gjpo.aspx

U.S. Environmental Protection Agency, Fact Sheet on Ocean Dumping of Radioactive Waste
Materials, Office of Radiation Programs, Washington, DC, 1980 (available at
www.EPA.gov).

U.S. Environmental Protection Agency, Data from Studies of Previous Radioactive Waste
Disposal in Massachusetts Bay, Office of Radiation Programs, 520/1-84-031
Washington, DC, 1984 (PB85 170 066/AS).

U.S. Environmental Protection Agency, EPA 402-R-05-007, Technologically Enhanced
Naturally-Occurring Radioactive Materials from Uranium Mining, Volume 1: Mining and
Reclamation Background, Radiation Protection Division, Washington, DC, June 2006
updated June 2007.

U.S. Environmental Protection Agency, EPA 402-R-05-008, Technologically Enhanced Naturally
Occurring Radioactive Materials from Uranium Mining, Volume 2: Investigation of
Potential Health, Geographic, and Environmental Issues of Abandoned Uranium Mines,
Radiation Protection Division, Washington, DC, April 2008.

U.S. Environmental Protection Agency, EPA 402-R-05-009, Uranium Location Database
Compilation, Radiation Protection Division, Washington, DC, August 2006.

207

U.S. Fourth National Report-Joint Convention on the Safety of Spent Fuel Management and on the Safety of Radioactive Waste Management

www.ingramcontent.com/pod-product-compliance
Lightning Source LLC
Chambersburg PA
CBHW081116170526

45165CB00008B/2456

* 9 7 8 1 4 8 1 1 4 2 8 2 3 *